- 2005年韩国坡州东亚书籍设计论坛／2005 Interchange Symposium of East Asia Book
- 2005年杉浦康平在韩国坡州东亚书籍设计论坛期间／Sugiura Kohei in 2005 Interchange Symposium of East Asia Book.

长期以来的装帧观念只停留在二维平面构成和外在打扮的层面，这使得书籍设计的认知范围相对狭窄，影响设计者努力就文本进行有创造性的设计。我们提出书籍整体设计（Book Design）的概念：完整的书籍设计要求设计师完成装帧（Book Binding）、编排设计（Typography Design）和编辑设计（Editorial Design）三个层面的工作，即"Book Design"是一个可视化信息再造的过程。

For many years, the meaning of book binding has stayed at two-dimensional formation and outer decoration. This narrow perception has been unchanged. Designers have not made due efforts to make creative designs for texts. According to the holistic book design concept, a perfect book design is where a designer completes three processes, book-binding, typography and editorial design. Book design is an activity that visualizes information that already exists.

◆ 2012年在德国奥芬巴赫柯林斯勒博物馆举办个展／Delikat:
Lu Jingren Book Design Exhibition in Klingspor Museum,
Offenbach, 2012

◆ 2012年在德国奥芬巴赫艺术设计学院教学／Workshop in HFG University of Art and Design, 2012

◆ 2012年在德国奥芬巴赫柯林斯勃博物馆举办个展／Delikat: Lu Jingren Book Design Exhibition in Klingspor Museum, Offenbach, 2012

Imitating and Innovating:
Book Design by Lu Jingren

传承·创造

法古创新
敬人人敬

上海人民美术出版社

◆ 敬人设计工作室／Jingren Art Design Studio

序 1

与吕敬人先生相识，是因为他是上海人民美术出版社著名连环画大师贺友直先生的弟子，他每次来沪，常会邀我一起看望其恩师。而熟识吕敬人先生，是因为他每次为贺老做书时透出的对中国文化的真情和对书籍设计事业的热爱，让我深受感动。

吕敬人，出生于沪上书香之家，历经知青十年艰苦岁月，从一名出版社的编辑出发到今天成为中国书籍设计界的领军人物，他的影响力足够改变中国图书的时代面貌。其实，作为和他交往颇多的出版业同道者，我感受更深刻的是他的文化担当和情怀。40年来，他始终视传承中华书卷文化为己任，不为名利，不畏逆境，不惧非难，以超出常人的坚韧力量，扛起中国书籍设计发展的责任。虽然他淡淡地称之为"修行"，然而我们都发自内心地钦佩他。当然，他更是一位踏踏实实工作在第一线的设计家，从未停下过自己的脚步。

本书记录下了他的优秀作品和足迹，而我们尤为看重的是他在中国首先完整地提出了"书籍设计"的理念。"装帧设计"和"书籍设计"虽只是一词之异，实则是一次根本性的变革。他对这一理念的诠释、实践和推广，极大地推动了中国图书的进步，也大大提升和转变了美术编辑在本行业的位置和角色。当今，越来越多的书籍设计人才脱颖而出，几乎全都得益于吕敬人先生为他们指明的方向。

现在，当有人对传统图书在传媒业的地位产生怀疑时，有吕敬人先生这样的引领者，我们就应该有更坚定的信心，中国图书事业的明天会更美丽。

是为序。

李新
原上海人民美术出版社社长
2017年9月

Foreword 1

I know Mr. Lu Jingren because he is a pupil of He Youzhi, a famous comic strip artist of Shanghai People's Fine Arts Publishing House. And he always invites me to visit his teacher when he arrives in Shanghai. I am well acquainted with Lu Jingren because the books he designs for Mr. He reveal the true feelings of Chinese culture and the love of book design, which moves me a lot.

Born to an intellectual family in Shanghai, Lu Jingren becomes the leading figure of book design in China from an editor of a publishing house after many years of painstaking efforts. His influence is strong enough to change the face of Chinese books. In fact, as a fellow of the publishing house, what impress me most is his cultural commitment and feelings. Over the past 40 years, he has always regarded the inheritance of Chinese book culture as his duty. He is not working for fame or fortune, and without afraid of hardship and difficulties. And he has borne the responsibility of the development of Chinese book design with resilience. Though he calls it as "discipline", we all admire him from the heart. Of course, he is also a designer who works in the first line and never stops.

This book records his outstanding works and footprints. And what we are particularly interested in is that he is the first person to showcase the concept of "book design" in China. The "graphic design" and "book design" are only slightly different, but they represent a fundamental change. His interpretation, practice and promotion towards this concept greatly promote the progress of Chinese books, and also greatly elevate and transform the role and position of art editor in this industry. Nowadays, more and more book designers stand out, who should thank Lu Jingren who has led them direction.

There are doubts about the status of traditional books in the publishing industry today, but we should believe more firmly that Chinese cause of book publishing will have brighter prospect thanks to such leaders as Lu Jingren.

Li Xin
Former President of
Shanghai People's Fine Arts Publishing House
Sep.7, 2017

序 2

以传统的称谓，吕敬人先生是以七十古稀之年举办"书艺问道——吕敬人书籍设计40年"的回顾展。若论从事某种职业、专业、事业，"青春作赋，皓首穷经"的40年付出，应是抵达峰巅境界，收获成就的时刻。有这么多的单位、部门和友朋同道毫不吝惜地给他赞誉之辞，乐意为他策划筹办，在他，是值得为之自豪、欣慰的乐事；在我们，则是感受鼓舞和策励的快事。

"书"者，记录思想、传播理念的文字载体，"书籍"者，思想灵魂的延展并得以传承的外化形式，书籍设计在当下已成为越来越注重生活质量和新知学习的国人，亲近、识读、珍爱书的文质和文本艺术收藏价值的重要学问，成为改变和充实现代中国人内美、气质和审美修养的元素之一。

展览的策划者为吕敬人设计艺术给出了充分的评价："吕敬人是享誉海内外的中国书籍设计界的代表人物，是当代中国书籍设计观念创新的领军者，他的影响力已不局限于书籍设计界，对中国出版业、设计教育界、信息传媒、文化创意领域、国际交流等诸多方面产生的大影响，是新中国成立以来，弘扬中华书卷文化和推动中国书籍艺术走向世界的积极开拓者。"吕敬人作品获奖无数，荣誉称号众多：他主持策划了多届全国书籍艺术设计大展和学术论坛，以及各类专题展览、学术交流；创办主编了第一本专业学术刊物；潜心致力于设计艺术教学和人才培养工作；他满世界地讲课，举办交流展，编写出版专著；不遗余力地将中国书籍艺术推广到世界各地……当敬人将智慧和生命的最好年华义无反顾地奉献给了他挚爱的事业时，没有什么比得上人们的认可，更能让他感受到强烈的存在感了。

以一个圈外人的眼光，透过眩目的光环看吕敬人的建树，是建立在他数十年实践与研究基础上的关于"书籍设计""编辑设计""信息视觉化设计""艺术×工学＝设计^2""传统设计现代语境""书之五感""设计要物有所值""书籍设计须触类旁通""纸有生命""书筑"等理念创新，以及对文字内容以外——"书"的设计的前世、今生、来世令人信服的继承、开拓、创意、独造。如果说这个时代需要出"高峰"作品，出"大师级艺术"，那么吕敬人应当是该领域具备了一切品质的人选之一。

倏忽，吕敬人先生告知我要离岗退休了，人生到了可以自由主宰自己时间的阶段。但是有一份目标追求的人，既无退亦无休。至于岗位，只是传播艺术和理想的平台，对于有终身学习和工作需要以填充生命的人而言，吕敬人无非是换了一种方式而已。

巧的是，我和他曾在清华美院共事，虽说隔着不同的专业，但吕敬人全情投入书籍设计教学，积极开设新课，撰写多部教材，悉心教诲辅导学生，以及开办各类研究班等社会教育，赴各国大学授课广受好评的业绩时有耳闻。也因此得知他在校园内、师友徒儿中有着"地中海发型""万有引力白须"和"笑脸爷爷"的形象爱称。

也曾经，我和吕敬人还是当年下乡东北的"荒友"，有过同样的经历，今天这些"血色苍茫"的历练成为我们生命中值得珍视的一部分。我们有着同样的理想，若因能力有限，人生只能做小事，那就尽己所能做到最好；或因岁月苦短，人生只能做一件事，那就竭尽全力去争取做好，于是，生活和生命才有了我们沉醉其中的所谓价值和意义。

吕敬人就是这样一位值得我们尊敬、学生爱戴的人。

真诚地祝贺他，并且期待着他从业50年，甚或60年的艺术回顾展。

冯远

画家、中国文艺联合会副主席、中国美术家协会副主席、
中国文史研究馆副馆长、清华大学艺术博物馆馆长
丁酉年白露写于京西六和堂

Foreword 2

With the traditional title, Mr. Lv jingren held a retrospective exhibition of "Seeking Enlightenment for Book Art - 40 Years Development of Book Design" at the age of seventy. When it comes to undertaking a profession, specialty and career, long-time and arduous work for forty years should lead to the moment of harvesting fruitful achievements. Many organizations, departments and friends offer him the words of praise without hesitation and are willing to plan and make arrangements for him. For him, it is a pleasure worthy of pride and gratification; and for us, it is an encouraging and inspiring pleasure.

"Book" is the writing carrier of recording thoughts and spreading ideas, and "books" are the extension of thoughts and soul and the external form to be inherited. At present, book design has become important knowledge of Chinese, who pay more attention to the quality of life and are willing to acquiring new knowledge, reading and cherishing the form and content of books and the collection value of text art. It has become one of the elements of changing and enriching the inner beauty, temperament and aesthetic accomplishment of modern Chinese.

Planners of the exhibition have provided a comprehensive evaluation of Lv Jingren's design art: "Lv Jingren is a world-famous representative of the Chinese book design circle, and is the contemporary leader of innovative Chinese book design concept; he has great influence on the Chinese publishing industry, design education circle, information media, cultural and creative fields, international exchanges, etc., in addition to the book design circle; he is also an active pioneer to carry forward Chinese book culture and to promote Chinese book art to the world since the founding of new China." Lv Jingren's works won numerous awards with many honorary titles: he presided over and planned many of national book art exhibitions and academic forums, as well as various special exhibitions and academic exchanges; he started and supervised the publication of the first professional academic journal; he has been devoting himself to the design art teaching and talent training; he lectured around the world, held exchange exhibitions and compiled and published monographs; he spared no effort to promote the Chinese book art to the world ... When Mr. Lv dedicates his wisdom and the best time of his life to his beloved career without hesitation, nothing can make him feel more strong sense of presence than winning people's recognition.

To appreciate Lv Jingren's achievements from the perspective of an outsider and through all his glorious titles is based on his decades of practice and research on innovative ideas such as the "book design", "editorial design", "information visualization design", "art x scheme = design^2", "traditional design and modern context", "five senses of books", "best value for design", "comprehending book design by analogy", and "paper has life", "book building", etc. as well as the convincing inheritance, development, originality and independent creation of the past, present and future development of "book" design beyond the content of the text. If the current age is in need of "outstanding" works and "master art", then Lv Jingren should be one of the candidates having all qualities in this field.

Suddenly, Mr. Lv Jingren told me that he was to retire, and his life reaches a stage that he can be free to arrange his own time. But for a person who pursues his goals, he is never retired. The post is just a platform to spread art and ideal. For a person who desires lifelong learning and work to enrich his life, Lv Jingren just changes the model of learning and working.

Coincidentally, both of us have worked in the Academy of Arts & Design, Tsinghua University. Although we engaged in different disciplines, I often heard his well-received performance of throwing himself wholeheartedly into the book design teaching, setting up new courses, writing a number of teaching materials, seriously teaching and tutoring students, running social education such as various tutorial classes, etc. and lecturing at universities of many countries. I also learned that he has vivid pet names on campus and among colleagues and students such as "Mediterranean hairstyle", "gravitational white beard" and "grandfather" with smiling face.

Lv Jingren and I once were the "friends in the Great Northern Wilderness" when we worked in the countryside of northeast China in those years; we have the same experience, and the arduous work become part of our life worth cherishing today. We have the same ideal, which is that if our ability is so limited that we can only make minor achievements in our lives, the we should exert our utmost efforts for the best; or because life is short, if we can do only one thing in our lives, then we will make every effort to fight for a good cause; life is thus endowed with the so-called value and significance for which we indulge in.

Lv Jingren is such a person worthy of our respect and students' love.
I sincerely congratulate him, and look forward to retrospective exhibition for the art that he has devoted fifty or even sixty years of efforts to.

Feng Yuan

Painter Vice Chairman of China Federation of Literary and Art Circles, Vice Chairman of Chinese Artists Association, Deputy Director of China Research Institute of Culture and History, Director of Art Museum of Tsinghua University Wrote in Luk Wo Tong in the western Beijing on September 7, 2017

序 3

我羡慕所有由敬人先生设计的书的作者。

我所写的书和所编的书,陆陆续续也有不少,但是还没有一本是"吕敬人"设计的,当然,不是敬人先生不给我设计,而是我不能轻易张口,我总想等有好书了再说。

事实上,从20世纪80年代他在中国青年出版社工作,到清华大学美术学院成为同事,算来我们也已认识了30年,我书房有许多他的作品,从他身上见证了中国书籍艺术设计的曲折和巨大进步,用一句感性些的话来说,我与敬人先生相交,是从一位虽然比我年长,但常以设计界同道和朋友的"老吕""敬人兄",到不经意间,突然意识到他已成为"书籍设计大家"的历程。

这是一种同在历史现场的同行对同行的评价和致敬,可想而知,我词语"冰山"之下的内容该有多么丰富!在"文化大革命"十年之后的设计复兴浪潮中,平面设计的"白马"始终勇立潮头,而集编辑设计、图文编排、文字传达和插图表现等为一身的书籍设计,仿佛是那白马头上的冠冕,耀眼生辉,在这一"冠冕"的群体中,那中间的核心人物,就是吕敬人。

一本书的设计,究竟该如何把握?是图解,是哗众取宠的推介,还是创造性的延伸?没有答案,这是真的,我曾在接受诗人兼书籍设计师的曹辛之(杭约赫)先生的赠书《最初的蜜》时请教过他,老人指着诗集封面——设计成雕版风格的一行字慢慢念道:"领你去会见自己。"

我想,敬人先生之所以成就斐然,真相可能就在于此,他的设计艺术立场,既非向内容投降,也非主观着去超越,更不是市场的迎合,而是以他的智慧,引领读者去"会见自己"。

"引领"是设计之技,而"会见"是艺术,能够会见"自己",这是"之道"。这"自己"既是文本作者的"自己",也是"设计师"吕敬人的自己,同时凝聚出"会见"后读者的自己,这些"自己",在书籍的传播和阅读中,构成了中国的"自己",亚洲的"自己"。

这次"书艺问道——吕敬人书籍设计40年"活动,策划者用了一个很贴切的主题词:"不摹古却饱浸东方品位,不拟洋又焕发时代精神。"这正是敬人先生的艺术和设计追求的写照,蕴含着中国当代设计思想的大意味。我记起数年前与杉浦康平先生、敬人先生的约定,要与他们两位做一次亚洲文化与设计的对谈,现在,约定虽然因种种原因推后了,但关于文化与设计,敬人先生以他的实践,已经做了最佳的回答。

谨致由表祝贺。

杭间
艺术史学家、中国美术学院副院长、
中国美术学院美术馆馆长、
中国美术家协会理事理论委员会副主任
2017年8月31日写于杭州湖畔

Foreword 3

I admire all the authors whose books are designed by Mr. Lv jingren.

I have written or edited a number of books, but none of them was designed by "Lv Jingren". Of course, it was not because Mr. Jingren refused to design for me, instead, I have been thinking that I should ask for such a favor when I am about to publish a really good book.

In fact, he started to work at China Youth Publishing Group in 1980s, and then we became colleagues at the Academy of Arts & Design, Tsinghua University. It has been thirty years since we knew each other, and I have the collection of many of his works in my study room. From my contact with Mr. Jingren, I have also witnessed the huge progress as well as the twists and turns in the development of artistic book design in China. Mr. Jingren, who was once a fellow and friend of mine specialized in design, now has grown into "a master of book design".

My comment on him represents an observation and a respect from the perspective of a peer, so it is easy to find how rich the content below my word "iceberg" is! In the wave of design renaissance after the 10-year Cultural Revolution, the "white horse" of graphic design has been a pioneer, while the book design that integrates editing and design, text-image layout, texts and illustrations seems to be a dazzling crown on the white horse, and the core figure in "crown" is exactly Mr. Lv Jingren.

How should a book be designed on earth? Should it be designed for illustration, for a promotion clamoring for public attention, or for a creative development? Indeed, there is no answer to this question. I have asked Mr. Cao Xingzhi (Hang Yuehe), a poet and book designer, on this issue when he gave his book Initial Sweetness (《最初的蜜》) to me as a gift, then he slowly read the engraving-like designed line of words on the cover of the poetry – "lead you to meet yourself".

I believe Mr. Jingren's prominent achievements are largely attributed to his stance of design art – leading readers to "meet themselves" with his wisdom, rather than giving way to the content, nor surpassing the content subjectively or pandering to the market.

"To lead" refers to the design technique, while "to meet" is an art, and to meet "oneself" is the path. The "oneself" can either by the author "himself/herself", or the "designer" Mr. Jingren himself, or the readers themselves after reading the book. These "oneselves" have formed Chinese "themselves" and even Asian "themselves" in the process of spreading these books.

The planner used a precise expression to describe the theme of the activity - "Tao of Book Design—A 40 years journey of Book Design by Lv jingren", namely "to achieve oriental style without drawing directly from the past, to capture modern spirit without mimicking the West". This exactly reflects Mr. Jingren's art and design pursuit, embodying the contemporary design thought of China. I remember years ago, Mr. Kohei Sugiura, Mr. Jingren and I have agreed to hold a discussion on the Asian culture and design. Although our agreement has been delayed for various reasons, Mr. Jingren has already offered the best answer to the development of culture and design with his practice.

Hereby I would like to extend my sincere congratulations to this occasion.

Hang Jian

Art Historian, Vice President of China Academy of Art,
Director of Art Museum of China Academy of Art, Director
of Chinese Artists Association, Deputy Director of Academic
Committee of Chinese Artists Association
Beside the lakeside, Hangzhou, on Aug. 31, 2017

序 4

弥漫着东方的清馨、美不胜收的书籍艺术之花,集吕敬人书籍设计之大成的"书艺问道"展,终于盛开在中国首都北京。值此展会开幕之际,我愿从隔海相望的东京,亦是吕敬人的第二故乡送上由衷的祝福。

本次展览,有以下几个看点。

● 第一、"汉字文化"在21世纪面临新的挑战。中国独创的千姿百态的汉字表现,又以重构一种灵奇震撼力的姿态而展开。

源于横竖自如、"天圆地方"的格子结构瞬息幻化成自由奔放的中国"书法"具有乐趣的表现。吕敬人从文字到书法、从矩阵到掀起涡旋般的字体群,对文字注入"气"的跃动与生机。

● 第二、对柔软而坚韧的中国"纸"文化的重新审视。

世界上最早掀起手抄纸革命的中国,今天再度焕发出强大生命力,让吕敬人做的书别具风情。纸张千变万化的手感、沁人心脾的气息为书增添华彩,催生人们每翻一页都不由引发对山川草木的乡愁。

● 第三、中国传统工艺的精湛使"穷尽手上功夫的造本术"的复活。

吕敬人和中国的同道们,不满足于现代机械印装书籍(折页、装订、裁切)立方体的批量生产,回归手工做书:触摸像鸟的羽毛般的丝绢质感,柔韧又饱孕体温,可谓专注中国特色的手工技艺。其丰穰的创想,为中国许多艺术家提供了爽快淋漓的知性的畅游平台。

● 第四、"创造围绕书籍的人的连环"。

一本书,得到志同道合者的参与,产生巨大的圆环,将著者—编者—出版社—设计师—印装企业—书店—读者有机地联系在一起。圆环,通过人们的齐心协力,环环相扣而生成。吕敬人在这个圆环中,他让年轻人发现做书的有趣,并附加了培养接班人的教育平台。他机智幽默、古道热肠的为人,不知疲倦的活力,促进这个圆环不断地扩大,在尊重和友情基础上,纷扬散乱的世界被凝聚打造成连带的圆环。

静心细加观看，在本次展会上，还可以不断发现吕敬人开展书籍革新的无数独创性尝试。

抚今追昔，我与吕敬人的邂逅，第一次是在近30年前的1989年，第二次是1992年。两次都为期一年，目的明确——以我的工作室为立足点从事出版（书籍设计）的学习。当时我（和杉浦工作室的同仁）正值20世纪七八十年代长达20年轰轰烈烈的"杉浦书籍设计语法集大成时期"，忙得不可开交。

然而，全体同仁都为吕敬人的人品——和颜温厚、不辞辛苦和废寝忘食的拼搏精神所感动，大家都愿意接纳他自由参加工作室的工作。我也每每为吕敬人超强的吸收力、创想力而感叹。深感我期待的"东方书籍设计"理想实现有望，因为最强大的生力军就在眼前！

吕敬人回国后，与突飞猛进的中国出版业同步成长，积极吸收中国传统的书籍艺术和工艺之精华，创作了追求中国艺术核心的精湛书籍设计。他的设计气韵生动，饱含东方的自然观与审美意识。通过温情回荡的设计力与颇具说服力的论考，开拓了中国年轻一代设计师的视野。

今天，吕敬人促成的中国书籍设计观念革新的背景，重合叠印着我与吕敬人在东京反复探讨的"东方设计语法的确立与实践"这一主题。

此处不遑详述，大抵如"21世纪的书籍设计——不是将东西方的思维方式割裂对立，而是互相尊重各自理念，兼容并包，形成'一'体"，"衍生形成'二而一，一而多'的书籍设计语法"……之类的命题。这是一种向对方靠近的看世界的方式，中国古人创造的"太极图"造型即为明证。

正是吕敬人，以更具活力的创意继承了杉浦书籍设计语法并发扬光大，从日本推向中国。蕴含"二即一，一即多"理念的书籍设计，以他的力量在中国的广袤大地，乃至世界得到远播，开花结果。

得此正统传人，我和支持我工作的广大同仁何等荣幸！谨对"天赐的礼物"——与吕敬人的邂逅，献上感谢。

杉浦康平

日本著名书籍设计家、设计教育家、
视觉信息设计家、亚洲图形研究学者
2017年10月写于东京（杨晶译）

Foreword 4

A truly beautiful flower of book art permeated with oriental delicate fragrance. The Exhibition of "Seeking Enlightenment for Book Art" that manifests the great achievements of Lv Jingren in book design is ultimately flourishing in Beijing, capital of China. On the occasion of the opening of the Exhibition, I would like to send my sincere blessings from Tokyo, a city separated by sea and also the second hometown of Lv Jingren.

The Exhibition is featured by the following highlights.

Firstly, "Chinese character culture" is faced with new challenges in the 21st century. The creative Chinese character manifestation in thousands of postures is showcased by reconstituting a kind of flexible and surprise posture.

The free, bold and unrestrained Chinese "calligraphy" transformed in the twinkling of an eye, which is originated from free lattice structure defined by "fitting between square and circle ducts", has an interesting expression. From characters to cal-ligraphy, and from matrix to vortical typeface group, Lv Jingren has devoted his all en-ergies to inject "animate" vigor and vitality into characters.

Secondly, the Exhibition re-examines the soft yet tough Chinese "paper" cul-ture.

Today, China, a country initiating the earliest handmade paper revolution in the world, is irradiating strong vitality once again, and renders peculiar expression to the book made by Lv Jingren. With ever-changing hand feeling and refreshing smell, the paper with new texture adds resplendent color to the book, and attracts people to spontaneously have nostalgia towards mountains, rivers and vegetations when turning every page.

Thirdly, the exquisite traditional Chinese arts and crafts revive the "book-creating process that exhausts sleight of hand".

Lv Jingren and Chinese people engaging in the same pursuit are unsatisfied with the volume production of cubic books printed and bounded by modern machinery (folding, binding and cutting), and return to make books by handwork: with a texture like bird feather, the tiffany is flexible and has body temperature, and can be seen as a manual skill full of Chinese characteristics. The well-rounded creativity has provided numerous Chinese artists with pleasant and intellectual platform for free imagination.

Fourthly, the Exhibition "creates an interlink of people that centers on books".

A book, with the participation of like-minded people, can generate a huge ring that connects all stakeholders in an organic way, including the author, editor, publishing house, designer, printing and binding enterprise, bookstore and readers. Thanks to people's concerted efforts, the ring generates by linking one another together. In this ring, Lv Jingren allows young people to discover the interestingness of book-making and adds an educational platform for cultivating successors. His tactful, hu-morous, considerate and warmhearted behavior, and inexhaustible energy has

◆ 2017年在北京今日美术馆举办的"书艺问道——吕敬人书籍设计40年"展现场／Tao of Book Design: A 40 Years Journey of Book Design by Lu Jingren in Today Art Museum, Beijing, 2017

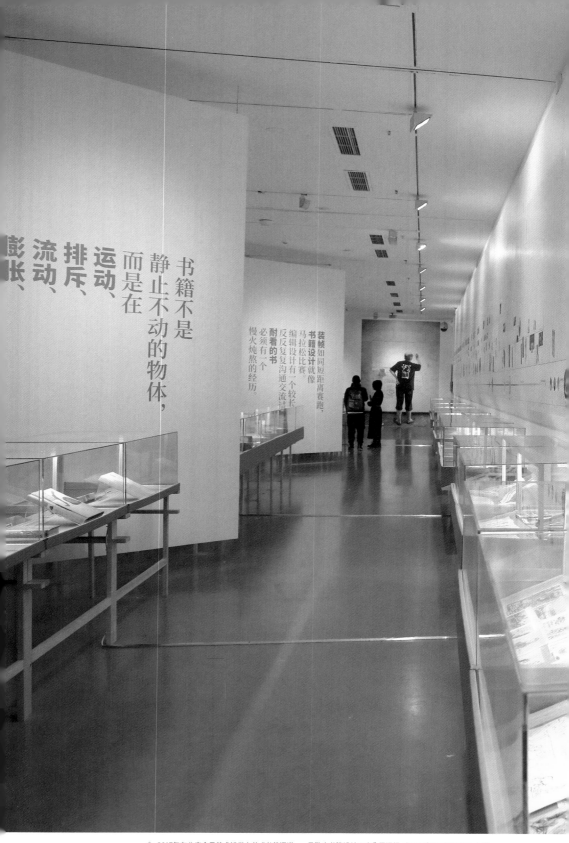

◆ 2017年在北京今日美术馆举办的"书艺问道——吕敬人书籍设计40年"展现场／Tao of Book Design: A 40 Years Journey of Book Design by Lu Jingren in Today Art Museum, Beijing, 2017

前言

我出生于20世纪40年代的上海,成长于天翻地覆社会变迁的时代,生活于充满激越动荡的政治漩涡中,亲身感受"文化大革命"与下乡改造的困惑和苦难,直到迷茫的30岁之时,"文化大革命"才结束。1978年,我踏进出版行业,艺术生命从零起步,走到70载古稀之年的我,似乎还有着当年的人生欲求。

这一生庆幸赶上中国改革开放的年代,比父辈、前辈、师长幸运得多,正当而立之年的我用加倍的努力追赶着失去的青春时光。旅途中有父母的训导,有恩师的指点,有同道的激励,有来自世界各方友朋的善待厚爱,使我的40年人生一直在从事着自己热爱的工作,我感到很幸福。

近40年改革开放,我亲身经历并目睹中国书籍设计观念的变化,其中包括年轻设计师的努力付出和迫切提升的学习欲望。我的作品并不是很优秀的,但我的设计经历或多或少可以映衬出中国书籍设计这几十年走过来的轨迹。

想象力来自深植于本土文化土壤的根所吸收的养分,但还需要适应时代的人文环境长出新的枝叶。复制传统只是一种仿效,法古要创新,传承亦开拓,设计的生命才能久远。我要特别感谢我的恩师杉浦康平先生,正是他的指点迷津,才使我有了这份直至今天的传播东方文化的坚持并成为现在的我。

我的十位弟子,也是与我共同成长的同道好友。我欣赏他们自始至终拥有做书的执著和热情,以及保持观念意识的不断更新,做到持之以恒,精益求精,厚积薄发。虽弟子,亦师亦友,本书也记载了他们与我相处的经历与说法。

读书是修行,做书是苦旅;
做书是一种责任,做好书也是一件善事;
美书,留住阅读。

吕敬人
2017年夏日

Introduction

Born in Shanghai in 1947, I experienced dramatic changes in my early days. I went through times of political instabilities and hardship, including the Cultural Revolution and relocation to remote areas. The nightmare of the revolution came to an end when I was 30. I was "reborn" in 1978 when I started my career in the publishing industry. Though I am 70 now, I still feel that I am passionate like I was 40.

I was more fortunate than the generation of my parents, teachers and mentors, because of the reform and opening-up of China. In my 30's, I made a decision to make up for the lost time. Thanks to the resolution, I could enjoy the second half of my life, doing what I wanted. Of course, it would not have been possible without coaching, lessons, encouragement and supports from my parents, teachers, colleagues and friends from home and abroad. They truly make me feel happy.

China began reforming itself and opened its door to the world, more than four decades ago. During the period, I personally witnessed changes of ideas in book design in my country. Take the efforts and desire of young designers as an example. I believe that my works can demonstrate the advancement of the Chinese book design over the last several decades, though they are not perfect. The book includes my 10 proteges. They are also my supporters who have let me along my career. I like their stubbornness to and passion for books; ever-changing ideas; relentless pursuit of perfection; and, humble attitudes to design.

My imagination originates from the depth of the Chinese culture. Yet I think that I have to be able to take advantage of shallower parts of it, like growing branches and leaves, instead of focusing on the trunk. A simple copy of a tradition is just an imitation. To constructively inherit a tradition, we have to create but also cultivate and pass it down to our children and their children. In this way, the vitality of book design would last long.

I want to extend heart-felt appreciation to my master Kohei Sugiura as he steered me in the right direction. Thanks to his coaching, I could continue with my work of promoting oriental culture and become who I am now.

Reading a book is training oneself, and making a book is
practicing penance. Making a book is responsibility, and making a good book is a good deed. A person should have a sense of shame, and a work should have a conscience.

I want to deliver my gratitude to Yi Ki-ung, CEO of Youlhwadang, for his unwavering support and attention. I also give my thanks to Kim Eon-ho, chairman of the Bookcity Culture Foundation, for inviting me to this exhibition; and also for his visit to Beijing so that he could share his idea about the subject of this exhibition with me and enthusiastic encouragement to me. My thanks also go to designer Ahn Sang-soo, Chung Byeong-gyu and Baek Geum-nam, other Korean friends and colleagues for their boost. I certainly believe that this event will go well as planned because of the support from the host, Pajubookcity; passionate planning and aids from Lee Hwan-gu, head of the secretariat, Kim Yeon-sook and Lee Ho-jin; and, efforts of my 10 proteges and others from my studio. I give my appreciation to them, as well.

Lu Jingren
Summer 2017

◆ 2016年"法古创新·敬人人敬 —— 传承创造：吕敬人的书籍设计与他的10位弟子展"在韩国坡州书城举办／Imitating and Innovating: Book Design by Lu Jingren and His 10 Proteges Exhibition in Paju Book City, Korea, 2016

不空谈形而上之大美，不小觑形而下之"小技"，东方与西方、过去与未来、传统与现代、艺术与技术均不可独舍一端，要明白融合的要义，这样才能产生出更具内涵的艺术张力，从而达到对东方传统书卷文化的继承拓展和对书籍艺术美学当代书韵的崇高追求。

I am not talking about metaphysical beauty. While paying attention even to "minor techniques" of physical things, we have to walk the fine line between the East and the West; the past and the future; tradition and modern; and arts and technology. Only when we comprehend the need of fusion and its meaning, the power of arts within book design can exert its strength. This materialized influence can expand the traditional book culture and pass it down to our children and their children, pursuing the noble aesthetic value of contemporary bookdesign.

Book Design 是令书籍载体兼具时间与空间、兼备造型与神态、兼容动与静的信息构筑艺术。书籍的设计与其他设计门类不同，它不是一个单独的个体，也不是一个平面，它具有多重性、互动性和时间性，即多个平面组合的近距离翻阅的形式。通过眼视、手翻、心读，全方位展示书籍的魅力。书籍给我们带来享受视觉、嗅觉、触觉、听觉、味觉五感之阅读愉悦的舞台。

Analogue media and e-books also have to follow this rule. Book design is a medium for a book to artistically build time and space; shape and attitude; and, dynamic and static information. Unlike any other designs, book design is neither simple nor flat. It is multiplex, mutual and diversely flat, which is influenced by time. Design is a form for short-distance reading. Through eyesight, hand-turning and reading by heart, overall charms of a book are felt. Books are a gift to us as they please our five senses: vision, smell, touch, hearing and taste.

目次
Table of Contents

- 013 ● **序1-4**
  Foreword 1-4
- 027 ● **前言**
  Introduction
- 039 ● **传承拓新　01**
  Imitating and Innovating
- 097　　　　● **中国当代的书籍设计**
  Contemporary Book Design in China
- 107 ● **聚珍汇集　02**
  Collect the Precious
- 125 ● **装帧入门　03**
  Entry into Book Binding / 1978–1989
- 143 ● **求学寻道　04**
  Seek Learning and Find a Way / 1989–1998
- 161　　　　● **影响我一生的两位恩师**
  Two Masters Who Gave Influence to My Life
- 173　　　　● **杉浦康平的公式：艺术×工学＝设计²**
  Kohei Sugiura's Formula: Art × Technology = Design²
- 185 ● **书籍设计　05**
  Book Design / 1998–
- 221　　　　● **装帧与书籍设计是折射时代阅读的一面镜子**
  Book Binding and Book Design is a Mirror That Reflects the Reading of an Era
- 259 ● **教学生涯　06**
  Teaching Career / 2002–
- 277　　　　● **书籍设计教育的学术主张**
  Academic Argument about Book Design Education
- 279　　　　● **关于网格设计的思考**
  Thoughts on Grid Design
- 284　　　　● **社会化设计教育的实验性探索：敬人书籍设计研究班**
  Explorative Study on Socialized Design Education: Jingren Book Design Workshop
- 291 ● **韬规家训　07**
  Lessons from My Parents / 1947–
- 297　　　　● **自述**
  Self-description
- 319 ● **同道弟子说　08**
  Pupils Walking with Me

Imitating and Innovating

传承拓新

韩国著名设计家安尚秀先生曾经说：『传统不只是过去的遗物，它是每个时代里最好的东西，在历史潮流的研磨中释放光芒，传承至今。』中国古代悠久的书籍设计艺术和丰厚的书卷文化，对当代中国书籍艺术的进步产生了重要的影响，要继承优秀的书卷传统，法古一定创新，不能停留在复制拷贝古籍的层面，要符合时代语境进行创造性的工作，让传统书籍这一纸媒艺术能一代一代传承下去。

我有幸参与了国家图书馆善本再造工程，体会到了东方古籍的艺术魅力，慢慢领会了东方艺术的理想：『悦目初不在色，盈耳初不在声。』★中国著名哲学家冯友兰语 艺术的审美境界不只体现在造物之外，更是浸润于内，富于暗示。『天时、地气、材美、工巧，合其四者然而可以为良。』★考工记 古人将艺术与技术，物质与精神之辩证关系进行如此精辟的阐述，即是形而上和形而下的完美融合，在这方面我想做一个东方书籍美学的寻梦者和实践者。

Famous Korean designer Ahn Sang-soo once said, "Tradition is not just relics of the past, but also something that can be cherished throughout generations. It is also something that is passed down from the past, while being modified and upheld through the tides of the history." The historical book design art and rich culture of China gave a critical influence to the advancement of contemporary book art in China. We have to pass an outstanding book tradition down to next generations, creating something new based on them. We should not end up in simply mimicking old books. Walking in the same pace of changes of the world, we must let books, a traditional paper medium art, make their way to the future, through creative efforts.

 I was lucky enough to participate in a national library's project that recreated old books. The participation offered me a chance to experience and to understand the artistic attractiveness of Eastern classics. Through the process, I gradually knew the meaning of Feng You-ran's comment. The famous Chinese philosopher once said about the ideal of Eastern arts: "What please the eyes does not lie in colors; what amuses the ears does not sit in sound." Aesthetic borders are not only for exploring creative works and understanding them, but for making them stay inside deeper and exist in implication.

"Weather, moisture on the ground, beauty of materials and sophisticated skills should go hand in hand for an excellent work," said Kaogonji, China's oldest craftsmanship book.

The wise who lived in the past were so insightful that they analyzed the relationship between arts and technology, and between spirit and substance, this way. They pursued perfection by fusing metaphysics and physics. Adding to that, I want to become a person who seeks one's dream and, at the same time, practices oneself.

001 S ⇨ 350 × 530 mm
D ⇨ 1999

朱熹榜书千字文

中国青年出版社

《朱熹榜书千字文》全书以原版复制宋代名家朱熹千字文，保持了拓片的原汁原味，寻得一种古朴的梵夹装书籍形态。三册合一帙，外函将一千个字反向雕刻在桐木板上，仿古代木雕印刷版，以皮带串联，如意木扣锁合，将东方书籍形态进行了创造性延展。封面以三种不同色彩的特种纸单色印刷，以中国笔画点、撇、捺作为上、中、下三册书的基本符号体现书法主题。书籍裸露锁线，既便于充分翻阅，又便于临帖。内文版式以文武线为框架，将传统格式加以演化，注入大小不同的文字字号，以及粗细截然不同的线条：上下粗线稳定狂散的墨迹，左右细线与奔放的书法字形成对比性的烘托，追求动与静的平衡和谐。

Zhu Xi's Bang Shu, the Thousand-Character Classic

China Youth Press

This book is a copy of Zhu Xi's version of the Thousand Character Classic, a famous literary legacy of the Song Dynasty. It is a rubbed copy, published in such a plain and simple way that Sanskrit mantras are laid out. This is a series book, comprised of three volumes, and thousands of characters are reversely inscribed on paulownia wood tablets pieces in such way that the ancient woodblock printing was done. They are bound together through leather thread, as if they were combined, and thus adding an element of creativity to the form of Estern books. Each of the three covers, made of special paper, is in different color, and characters on them are in one single color. The three major factors (*jeom, ppi-chim* and *pa-im*) in the Chinese calligraphy are used for basic signals for each of the three volumes. The marks make readers know that they are calligraphic books. Their back is in an exposed binding so that learners can easily open them flat and practice characters with ease. The way that contents are printed is largely based on the line of literary and martial arts, a modification of traditional methods. It adopted different sizes of character and easily-identifiable diverse stroke slimness. Vertical strokes that give a thick and stably spreading feeling contrast with horizontal lay-outs of characters written in slim yet powerful touches, striking a balance between the dynamic and the static.

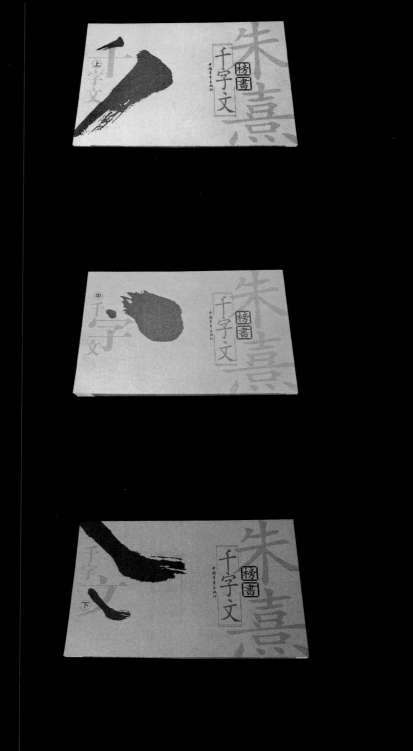

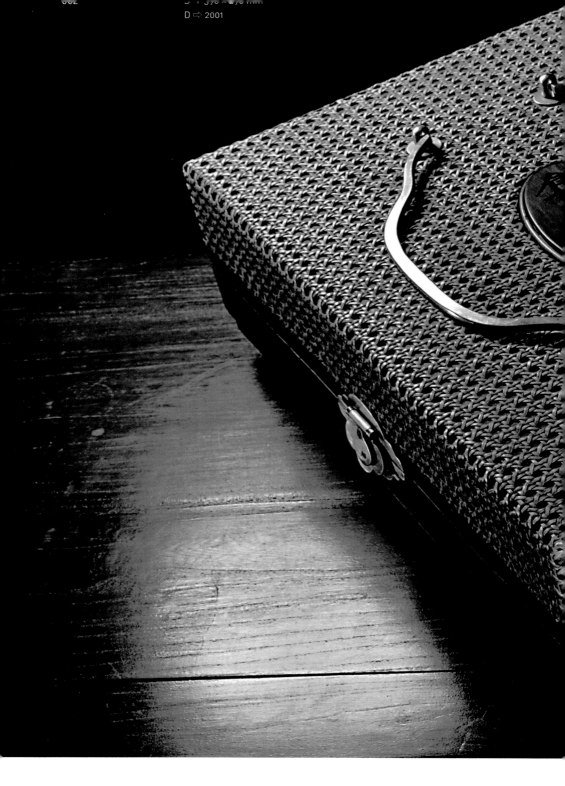

食物本草

北京图书馆出版社

Herbal foods

Beijing Library Press

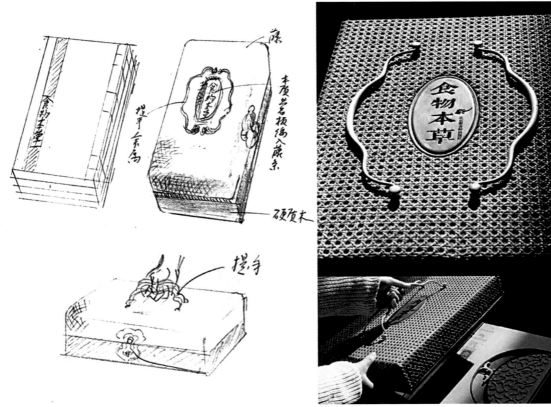

003   S ⇨ 260 × 50 mm
         D ⇨ 2002

**Bian Wang Lun
/ Shang Shu**

Beijing Library Press

尚书·辩忘论

北京图书馆出版社

004   S ⇨ 390 × 190 mm
       D ⇨ 2001

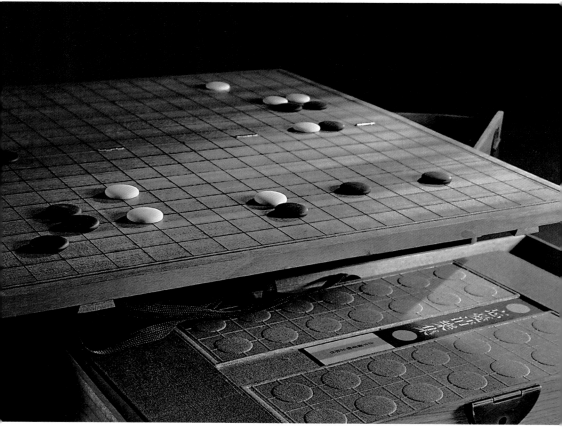

忘忧清乐集

北京图书馆出版社

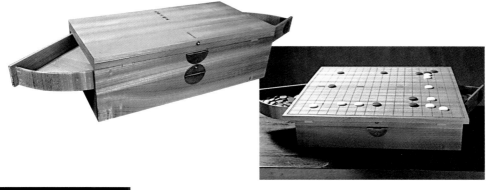
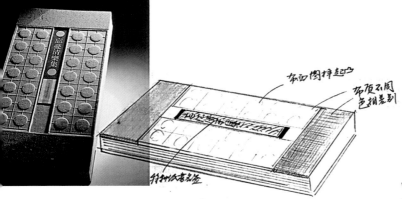

Wangyou Qingyue Ji

Beijing Library Press

005   S ⇨ 285 × 420 mm
        D ⇨ 2003

中国少林寺

中华书局

Chinese Shaolin
Temple

Zhonghua Book Company

S ⇨ 210 × 285 mm
D ⇨ 2001

赵氏孤儿

北京图书馆出版社

一本由 18 世纪风靡法国的法文《赵氏孤儿》剧本和明代雕版印刷本合而为一的书，作为国礼由国家领导人赠予法国总统。书做成双封面，中式封面篆刻为明版木雕文字，西式封面为法文译本，分别以装饰图形的格式呈现，一个是东方如意纹，一个是西方几何曲线纹。中国的雕版印刷和西方的铅字技术相结合，最终形成一个中西交融、中西互通，但又具两种语境的书籍形态。

The Orphan of Zhao Family

Beijing Library Press

This book is a combined version of *The Orphan of Zhao Family*, a French edition which went very popular during the 18th century in France, and another edition published during the Ming Dynasty. It was given as a gift of a Chinese President to a French President. Its double cover carries seal engravings in block letters of the Ming Dynasty in Chinese page and their translation into French. The first is in an Eastern pattern, ruyiwen, and the second in Western patterns with geometric curves. In addition, China's typesetting and the Western printing technology blend together for the publication to give a message of fusion. The book is also designed to have China and France communicate with each other and conserve their attractive linguistic environment.

007　　　S ⇨ 397 × 265 mm
　　　　　D ⇨ 2001

沈氏砚林
北京图书馆出版社

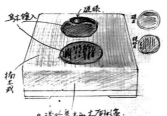

Shen Family's Inkstone Collection

Beijing Library Press

S ⇨ 290×190 mm
D ⇨ 2000

北京图书馆出版社

跋乾隆甲戌脂砚斋重评石头记

Postscripts to Zhi Yan Zhai Criticisms on "A Dream of Red Mansions"

Beijing Library Press

009　S ⇨ 210 × 285 mm　D ⇨ 2005

周作人俞平伯往来书札影真
北京图书馆出版社

文人手札，讲究书、画、印三位一体，书笺呈格式韵味，行文还经营布局。《周作人俞平伯往来书札影真》的作者是国学者，又为西学文化人，东西融合是该书设计风格的着眼点。将宣纸裁成边镶嵌入书函天地，形成传统线装书的柔软质感，启开书函是两本西式形态的书，体现作者中西合璧的学术特征。

Letters Between Zhou Zuoren & Yu Pingbo

Beijing Library Press

Letters Between Zhou Zuoren & Yu Pingbo, literati hands, is a trinity of exquisite calligraphy, painting, printing. Bookmarks are charming, and the layout is exquisite. Created by an author of both Chinese and Western culture, integration is the focus of the book design. Rice paper makes the soft texture of traditional thread-bound books, letter opening are two Western-style forms of book, reflecting the characteristics of Western and Chinese academic authors.

010

绘图五百罗汉详解 卷轴装

国家图书馆出版社

Understanding of
500 Paintings

National Library Press

S ⇨ 210 × 285 mm
D ⇨ 2005

国家图书馆出版社

绘图五百罗汉详解（函盒装）

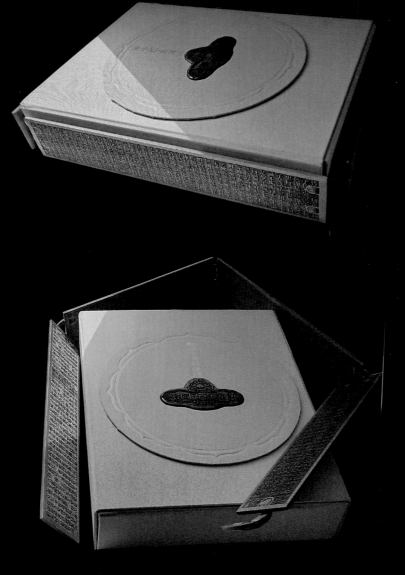

Understanding of 500 Paintings

National Library Press

011   S ⇨ 270 × 200 mm
D ⇨ 2002

证严法师佛典系列

静思文化志业有限公司

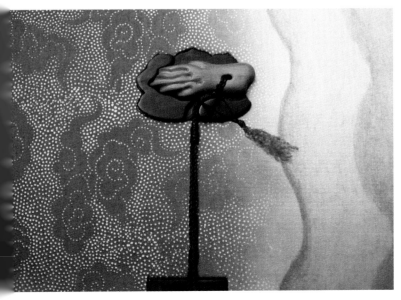

Sutra of Zheng Yan

Jingsi Culture Ltd. co

Picture of Miracle

National Library Press

013 S ⇨ 340 × 210 mm
D ⇨ 2001

奏鸣曲 为小提琴独奏和通奏低音而作

北京图书馆出版社

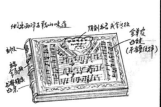
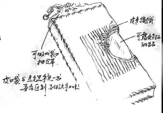

Sonate-A Violin Solo Col Basso

Beijing Library Press

014　　S ⇨ 320 × 430 mm
　　　　D ⇨ 2004

**Exhibition of Lost
National Treasure of
the Qing Dynasty**
―――
Zhonghua Book Company

清宫散佚国宝特集

中华书局

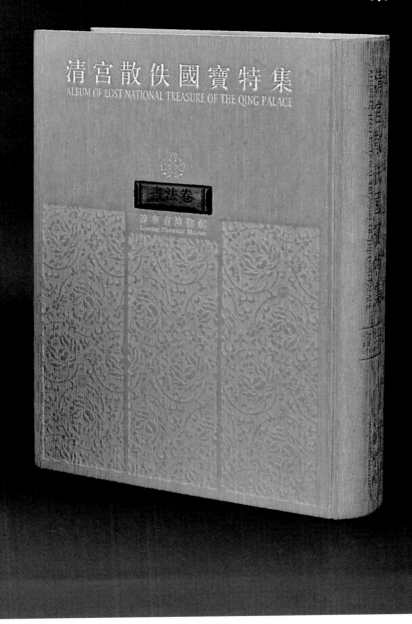

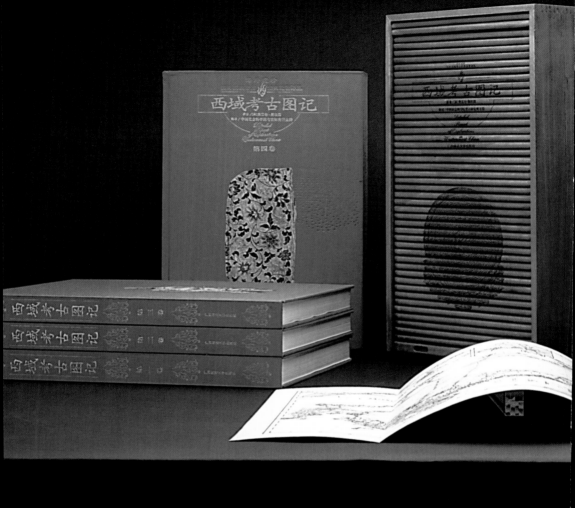

西域考古图记

广西师范大学出版社

Detailed Report of
Explorations in
Westernmost China

Guangxi Normal University
Press

The Great History of China

Cultural Relics Press

中国大史记
文物出版社

Imitating and Innovating

The Book of Tea and the Book of White Spirit

Beijing Library Press

S ⇨ 420 × 350 mm
D ⇨ 2001

北京民间生活百图

北京图书馆出版社

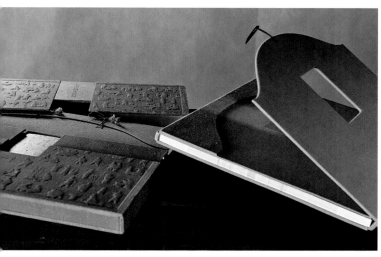

Folk Life of Old Beijing

Beijing Library Press

S ⇨ 240 × 360 mm
D ⇨ 2008

中国记忆——五千年文明瑰宝

文物出版社

《中国记忆——五千年文明瑰宝》以构筑浏览中国千年文化印象的博览"画廊"作为设计构想，设计的核心定位是体现东方文化审美价值。内文使用柔软的筒子页包背装结构，组成中国式阅读语境。每一部分薄纸隔页与正文内页的纸质形成对比，打造鲜明的触感层次。跨页采用 M 折法，使信息得到完整表达，并体现阅读的互动性。在编辑设计上，则强调文字与图像的主次关系和余白的节奏处理，并强调图像精美准确的印刷，使其具有逼真完善的还原度，增添了该书的欣赏性和学术价值。封面运用具有水墨意蕴的万里长城摄影作品为基调，书腰则以中国典型的文化遗产图像反印纸背，通过对折使视觉图形若虚若实、亦真亦幻，烘托出一种跨越中华历史时空的氛围。

《中国记忆——五千年文明瑰宝》是北京奥运会馈赠各国领导人的国礼书，设计力图要做到体现代表国家身份的文化气质，既彰显中国传统文化的典雅，又具有时代气息。该书于 2009 年获得莱比锡"世界最美的书"奖。

The Chinese Memory
-Treasures of 5000
Years of Civilization

Cultural Relics Press

The book is designed to give readers a sense of touring a gallery that displays thousand-year old exhibits of the Chinese culture. The core of the design lies in specifically exposing the value of aesthetic beauty of Eastern culture. The print is made

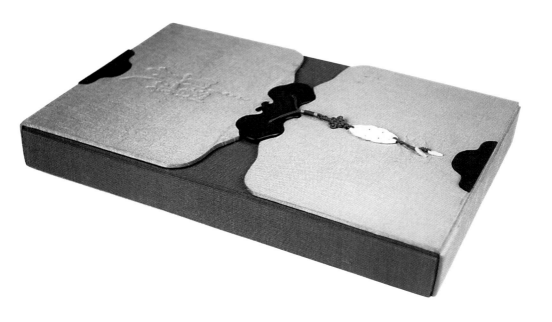

on smooth and thick paper, and its back is bound with thread and wrapped with big paper, helping readers conveniently enjoy it. Thin paper is inserted into every chapter, thus making a clear difference in touch between a regular printing paper and itself. The part where a reader touches to turn pages fully displays information and realizes interactive reading, by using an M-shaped folding method. The lay-out design underlines the relationship between characters and images and the processing of empty space rhythm. The delicate, accurate printing of the book makes images look real, thereby becoming the reason for appreciation and greater academic values. The image of the Great Wall appears on the cover in an implicit way of an ink-and-wash painting, and paintings of cultural heritage of China are on the backside of paper band. As such, those pictures are murky and give a sense of illusion that the Chinese history transcends time and space.

This book was given to many diplomatic delegations during Beijing Olympic Games. It is designed to give a feeling of representative of Chinese cultures, showing the elegance of China's traditional culture and keeping up with the flow of the time. The book was awarded "Best Books from all over the World" at the Leipzig Book Fair.

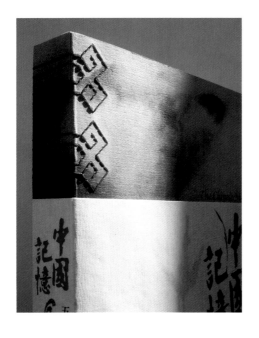

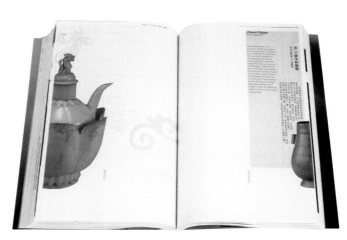

020

S ⇨ 245 × 380 mm
D ⇨ 2010

怀袖雅物——苏州折扇（精装）

上海书画出版社

《怀袖雅物——苏州折扇》是一套介绍世界非物质文化遗产中华传统扇子艺术的书籍。折扇被视为中国文人雅士的象征物，全书设计刻意透着中国的书卷风韵。与著者共同策划后，确定了本书设计以古线装、经折装、筒子页、六盒套等传统书籍形态作为基础，但又不拘泥于原有模式。设计理念强调主题陈述的时间概念，解读折扇工艺全过程。为让读者理解造就中国扇子之美的"天时、地气、材美、工巧"的人智物化的道理，强调文本资讯的编辑设计概念，关注视觉元素的采编、编辑、整合，运用多种现代书籍编排语言，有层次、有节奏、戏剧化地传达文化内涵。通过书页中的夹页、长短页、单拉页、对折页、M折页的翻阅设计，体现折扇的多层重叠特性。全书采用近十种纸，准确把握了纸张的语言和表情，让它们分别担当书中不同资讯所承载的角色。书脊订口特意为四册线装本分别设计梅、兰、竹、菊四君子的图案，装帧的缀订形式及函盒根据阅读本与珍藏本的不同用途，分别进行结构上的创新设计，新颖、质朴且庄重，以表现承其魂、拓其体的中国文化气质的追求。

本书获得

2010 年　第 4 届金光国际印艺大赛设计创意金奖　　　　2010 年　"中国最美的书"奖
2010 年　第 22 届香港印艺大赛冠军奖、全场大奖　　　　2011 年　第 62 届美国国际印艺大赛新颖书籍设计金奖

Elegant Article in Sleeve: The Folding Fan of Suzhou

Shanghai Fine Arts Publisher

The book Elegant Article in Sleeve is about the traditional Chinese folding fan which shows up on UNESCO's List of Intangible Cultural Heritage of Humanity. The book design had been created in cooperation of the authors. As fans are the symbol of Chinese literati and scholars, the book was intended to show the graceful and beautiful figures of Chinese books. This is why the book is based on forms of traditional books, including: seon-jang, a way of book-binding that puts a printed side outward; gyeong-jeop-jang, a method of bookmaking for the use of Buddhist monks; tong-ja-yeop, smooth and thick paper; and, yuk-hap-tu, a series of 6 volumes. But, it is not restricted to those elements. The design understands all processes of fan-folding craft, underscoring that themes are described in accordance with time concept. Authors made the book in such a way that readers grasp the beauty of making Chinese fans. The book includes: tian-shi, weather; di-qi, the moisture of the ground; cai-mei, fascinating materials; and, gong-qiao, sophisticated skills. It also focused on the collection, editing and arrangement of

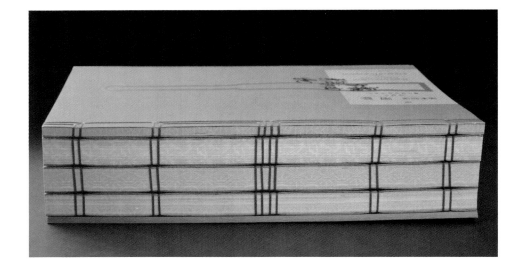

visual elements, while emphasizing the editorial design concept of textual information. In addition, it has diverse layers, by making use of typography methods of different modern books, and conveys rhythmical and dramatic textual information. It also has various ways of turning pages: inside pages in the middle of pages; inserted pages in diversified sizes; pop-ups; folding paper; and, M-shaped folding paper. It materializes diverse folding features of fan-making. More than 10 different kinds of paper were used in the book, each of which disseminates the information of their own feeling and texture. The backs of four seon-jang volumes are in patterns of four cardinal subjects - oriental orchids, bamboos, plum blossoms, and chrysanthemums. Ways of binding a book and making a book case are in structurally different designs for different purposes of a book: say, reading, or keeping on the bookshelf. The design concept of the book is to inherit the authentic atmosphere of the Chinese culture and, at the same time, pursue a higher dignity of it.

Awards

Gold Prize in Creative Design at the 4th Gold Mine International Printing Arts (2010)
Grand Prize at the 22nd Hong Kong Printing Arts (2010)
The Beauty of Books in China (2010)
Gold Prize in Fresh Book Design at the 62nd International Printing and Graphic Arts Exhibition of USA (2011)

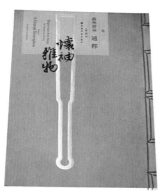

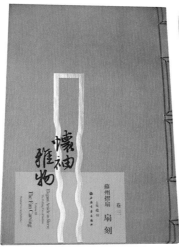

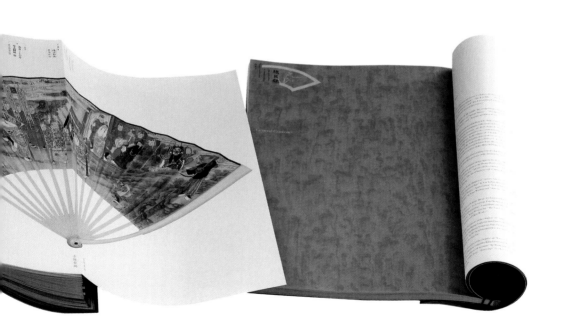

Imitating and Innovating

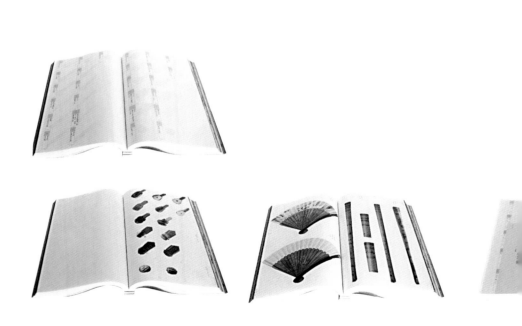

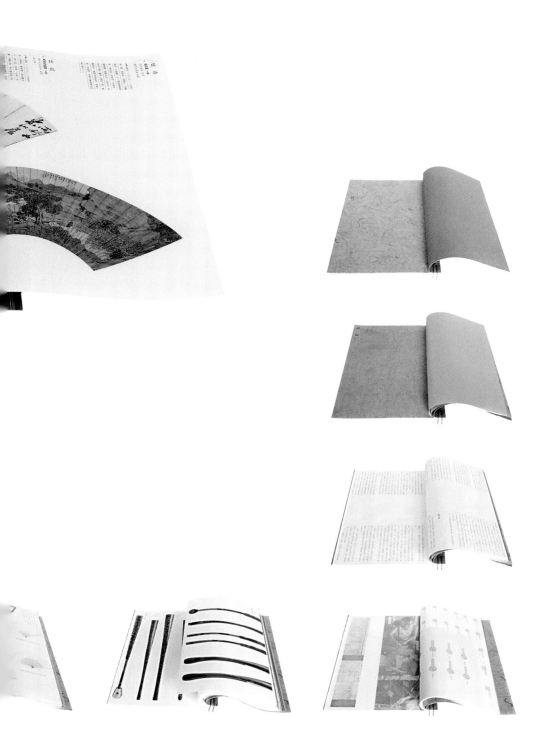

藏区民间所藏藏文珍稀文献丛刊精华本

四川民族出版社 + 光明日报出版社

**Rawwre And Ancient
Tibetan Texts
Collected In Tibetan
Regions Series
Collection Edition**

*Sichuan Nationalities
Publishing House +
Guangming Daily Press*

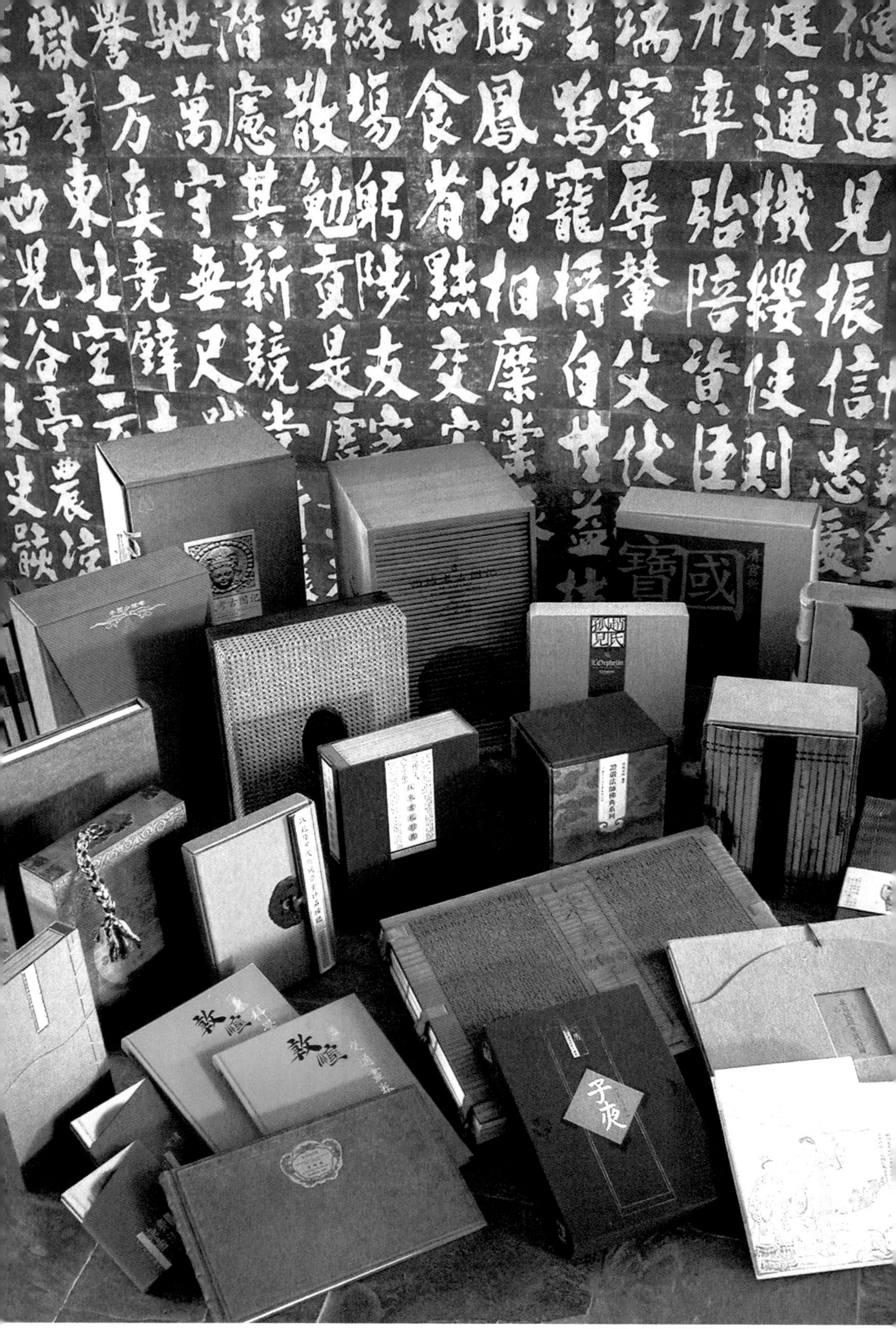

中国当代的书籍设计

中国的书籍设计在数千年漫长的古籍创造中经历着各种书籍制度的变迁，在不断完善中推陈出新，并不断衍生出新的书籍形态。中国近代书籍设计起步于20世纪初的辛亥革命，自此开启了一扇封闭已久的文化之门，开始受到外来多元文化的影响。在20世纪初的中国新文化运动中，鲁迅、丰子恺、孙福熙、司徒乔、闻一多、钱君匋等一大批文化人，演出者留学欧美日，把欧洲的各种流派的插图艺术风格和被日本称之为"装帧"的书籍设计引进中国，传统和外来文化融合，形成五彩纷呈的民国书籍艺术风景。1949年后，中国书籍艺术受当时苏联的现实主义影响，苏联、民主德国专家被聘请帮助提升印制技术，国家还派人赴东欧学习。那一时期最优秀的美术家们都服务于出版行业，如叶浅予、黄永玉、丁聪、曹辛之、黄胄、蔡亮、张慈中、任意、袁运甫等，涌现出一大批至今仍可称之为经典的装帧、插图之作。遗憾的是，20世纪60年代，中国社会、政治、经济动荡，直至"文化大革命"彻底摧残文化十年，中国的出版业发展停滞，书籍设计业陷入严冬低谷。1976年，"文化大革命"结束，冬去春来；1978年，改革开放，中国的书籍设计业真正迎来了艺术的春天。

20世纪，中国书籍制度经历了巨大变化，文本的竖排格式改为横排格式；繁体字变简体字；装帧手工工艺逐渐跨入大批量的工业化印制进程；维系了大半个世纪的活字凸版印刷，被90年代的平版胶印所替代，成为中国印刷的主流；被称为当代毕昇的王选开发了北大方正汉字数码排版系统，迎来21世纪电子数位化印刷的天下。当今，中国生产力、生产工具、生产关系的巨变，必然引发阅读载体、出版体系、产出授受关系、设计思维概念等革命性的范式转移。

90年代是中国改革开放的黄金期，思想精神的解放，使中国书籍文化丰富而多元，出版界书籍设计开始有了改变滞后观念的迫切感。出版、设计、印艺业的有识之士开启了广泛的国际化交流渠道，使更多的设计师们以开放的心态和学习的诚意，对东方与西方、传承与创新、民族化与国际化、传统工艺与现代科技有了新的认知。他们打破装帧的局限性，投入大量精力和心力，强化内外兼具的编辑设计，为创造阅读之美进行了有益的探索。很多年轻的设计师不拘泥于单一的体制环境，脱离国

◆《九叶集》／Jiu Ye

——《马克思画传》／The biography of Marx

◆《阿诗玛》／Ashma

◆ 《呐喊》/Na Han　　◆ 《彷徨》/indecision　　◆ 《欧洲大战与文学》/European War and literature

家体制，自主创业，以个体的设计人身份或独立工作室的多元模式，加入社会化的竞争。无论是体制内还是体制外，一批一批设计新人涌现，他们的优秀作品被读者喜爱，令国内外瞩目，因此才有了当今中国书籍设计的新的面貌。举一例，上海新闻出版局2003年开始主办"中国最美的书"评比活动，每届评出的20本书再参加莱比锡"世界最美的书"的评选。13年间，有287本书获得"中国最美的书"奖，其中的15本设计获得包括金、银、铜奖在内的"世界最美的书"称号。"中国最美的书"也成为书籍设计业的品牌，很多读者纷纷购买收藏，体现了设计的价值。

另外，自1959年开始的全国性的书籍艺术设计大展，虽跌跌撞撞经历时代的动荡和风雨，30年间只办了三届，而后17年已趋向稳定，1995年举办了第4届，1999年第5届、2004年第6届、2009年第7届、2013年第8届全国书籍艺术设计大展，其涉及面广，参与者多，是业内规模和影响力最大的赛事，其间还举办国际书籍设计论坛，极大地推动了书籍设计理念的发展。2006年，国家新闻出版总署设立了三年一届的国家政府奖（每届评出10本书）。以上三个大赛是出版行业内公认的国内三评奖活动，给年轻一代的书籍设计师带来设计竞争的机会和创作的动力。这些奖项也成就了设计师的事业。如吴勇、刘晓翔、小马哥+橙子、赵清、朱赢椿、何明、洪卫、杨林青、马仕睿、连杰+部凡、李让……他们在设计界不断崭露头角，使中国的当代书籍设计开始为国际关注。

中国书籍艺术在近些年虽有较大的发展，但还存在诸多问题。全国有约550家出版社和近万家杂志社，每年出版近30万种新书和10多万种再版书，还有大量刊物蜂拥入市。大批设计师被海量的设计和滞后的装帧观念拖累，部分设计师一人一年要设计400—500本书的封面，机械式的工作已使设计师失去创意的动力。一些出版单位为经济效益，只着力于书皮设计的表面功夫，为省时间和成本，放弃内在编辑力量的投入，并不断压低设计稿酬。浮躁的做事心态使不少书的文本叙述流于平庸，出现了山寨、模仿、不尊重版权等严重问题。低质导致许多出版物一面市就滞销，很快成为废品，许多书籍设计师因巨量劳动和低廉的设计费而无法生存。但这一现象并没阻止一批有良知的出版人和有责任感的书籍设计师开始反省做书的意义。他们不畏艰难，苦苦实践，使这一行业有了更多的共识：书不仅要有一件"漂亮"的外衣，还要有内在书籍设计整体概念的倾注。设计师应成为文本传达的参与者，像导演那样让资讯在页面空间中拥有时间流动的含义，使书成为文本诗意表现的舞台。出版不能只谈"价"而不顾"值"，要物有所值，原创的精品书能传世后代才体现做书的价值。设计是一种态度，设计要专一、有温度、讲细节，文本叙述的丰满才能带来阅读的动力。

当今，全球进入大资料时代，中国网民处于加速度生长期。据中国互联网络资讯中心统计，1997年中国上网户数仅62万，2015年中国网民达到6.84亿人，手机网民达6.2亿人，18年间增加1100倍。不能否认时下电子载体的盛行，给传统的出版业带来巨大冲击，许多出版机构都在分流精力和资金投入电子出版。显然21世纪的新技术带来的新的阅读载体会大量出现，这应该是件好事。不过近期也有一个现象值得我们留意，传统的阅读习惯并没有被年轻一代所放弃，好的阅读产品仍然会吸引他们。一些优秀的文化人纷纷成立个体文化出版或编辑策划公司，自筹资金与出版社合作出版符合市场需求也有阅读价值的书。这一出版模式，提供给设计师打破出版社固定思维和模式，发挥创意并设计出优质图书的可能性，而且这种例子越来越多。"中国最美的书"获奖者中很多是体制外的设计师。"世界最美的书"获奖者中自由设计师与体制内设计师的人数比例是10：3，前者比后者有更广阔的创作空间，可见不同的工作体制下创造的结果是有差异的。还有一个现象，一些不注重

◆ 部分获得"世界最美的书"奖的中国书籍设计／Some of the most beautiful books in the world. Design of Chinese book Honors

书店文化的国营书店在萧条，而私营的富有创意和具书卷文化的个性书店在兴起，并受到越来越多的年轻人的青睐。如北京的万圣书店、单向空间书店、库布里克书店，南京的先锋书店，上海的衡山和集书店，广州、成都、重庆的方所书店等。台湾的诚品书店已进驻苏州，上海浦西最高大楼——上海环球金融中心的62、63层和地下一层的诚品书店也将开业，"艺术＋阅读＋生活"已经形成当今书店的新模式。正因为虚拟的电子阅读的普及化，人们对于实体纸面书籍的好感度和需求也在与日俱增。一些具有个性的手工出版，或限量版书受到读者喜爱，市场上出现了许多引发关注的出版品牌，如香蕉鱼书店、加餐面包、暖书、连和部设计、友雅工作室……他们的书做得与众不同，精致到位，销量也节节攀升。由中国出版协会书籍设计艺委会和中国美术家协会平面设计艺委会联合举办的"全国大学生书籍设计大赛"已举办四届，推动了各艺术院校的书籍设计教学。书籍设计艺委会主编的《书籍设计》杂志，为完善学术研究，普及书籍美学，推广世界先进设计理念提供了很好的交流平台。中央美院徐冰策划的概念书展"钻石叶"广受欢迎，显现出人们对艺术图书的兴趣。吕敬人创办的"敬人纸语"书籍设计研究班，也是对传统阅读回归的顺应和新造书运动的提倡。

中国书籍艺术是一条动态发展的历史长河，中国的设计师要用敬畏之心珍惜祖先留下来的宝贵遗产，具有谦卑且冷静对照古人做书的进取意识，脚踏实地做好传承这门功课，同时又要有开放的胸怀，学习世界各国优秀文化，海纳百川，尊重发展规律，才能融入进步的潮流，这也是中国书籍艺术能持续发展的动力。传统不是模式化的复制，传承更不是招摇过市的口号，每个民族的设计不可能从自身传统文化的土壤中剥离出来，世界各国设计师在寻找现代语境下延展本土文化的新途径，这也应该成为当代中国设计师的理念追求。改革开放40多年，大批书籍设计师创作的优秀作品，像美书连接起来的一道彩虹，辉映出他们满怀热情，付出辛劳和智慧的做书心迹。

创造书籍之美，留住阅读。

◆ 部分获得"世界最美的书"奖的中国书籍设计／Some of the most beautiful books in the world. Design of Chinese book Honors

Contemporary Book Design in China

Over the Past millennia, books have undergone changes and achieved innovation through ceaseless reforms. Modern book design in China is said to have begun with the Chinese Revolution in the early 20th century. Those in the cultural and art communities back in the early 20th century, including, Lu Xun, Feng Zikai, Sun Fuxi, Si Tuqiao, Wen Yiduo and Qian Juntao, went overseas, to Europe, Japan and the US, and brought to China styles of illustration art of European art schools and book design which was called book binding in Japan. They have established a variety of the book design art culture, by fusing traditional and foreign elements. Influenced by the realist art of the Soviet Union since 1949, the Chinese book art upgraded its printing technology, by inviting experts from the communist union and East Germany. The Chinese government sent people to East European countries for further study. Exceptional artists of the time were engaged in the publishing industry, working hard on illustration and book binding which was still called classic. Examples are Xie Qianyu, Huan Yongyu, Ding Cong, Cao Xinzhi, Huan Zhou, Cai Liang, Zhang Zizhong, Ren Yi and Yuan Yunfu. Unfortunately, the socio-economic and political landscape of China in the 1960s was chaotic. The 10 years under the Cultural Revolution held up the development of the Chinese publishing industry, posing extreme difficulties to book design firms. Things have changed since 1976 when the revolution came to an end. It was more desirable that, in 1978, the government decided to reform and open up, as the book design industry finally entered the season of spring.

Book-related institutions in China went through huge changes in the 20th century: from the vertical writing to the horizontal one; from traditional Chinese characters to the simplified one; and, from manufacturing to mass production of book binding. Type-printing that had been used for half a century was replaced with offset-printing in the 1990s. Then, the country into digital printing ever since Wang Xuan, a modern Bi Sheng, invented the Beidafangsheng digital typesetting system (Bi Sheng 990–1051 AD was the inventor of type-printing.) Gigantic changes in productivity, production tools and production relation in China are making a revolutionary paradigm shift in reading equipment, formation of publishing system, relation of giving-and-taking products and ideological concept on design.

The 1990s was the prime time of the Chinese reform and opening-up. Liberation of thoughts became was mainstream and fruitful, led by the book culture in China. The publishing industry began to know the dire need of changing the unchanged perception on book design. The intellectual in the publishing, the design and the printing industry built a broad international exchange channel so that more designers had perception on the East and the West, nationalism and globalization, traditional craft and modern scientific technology, largely leveraging open-mind and passionate attitudes. They made all-out efforts to go beyond the limit of book binding and continued with the exploration in creating the beauty of book reading. Undeterred by the state system or other types of restrictions, young designers established their business as they found it pleasing to partake in the competition within their society, as an individual designer or independent workshop owner. China saw a wave of new designers coming, employed by national press corporations or self-employed. Their distinguished works were picked up by readers and drew attention from home and abroad, bringing a new aspect to the Chinese book design. For example, a Shanghai newspaper agency has held Most Beautiful Book in China since 2003. Every year, more than 20 books make their way to the competition Best Books from all over the World in Leipzig, Germany. For the last 13 years, a total of 287 were chosen as most beautiful in China. Particularly, the designs of 15 books earned the acclaim of gold, silver and bronze award. The title Most Beautiful Book in China is now perceived as a brand, and we can easily know the value of its design, as many readers buy it for keeping on their bookshelf.

The National Book Art Design Exhibition was launched in 1959, but held just three times during the first three decades, affected by chaos and troubles of the times. Yet, for the following 17 years, it was resume and held diverse awards. Examples are the 4th in 1995; the 5th in 1999; the 6th in 2004; the 7th in 2009; and, the 8th in 2013. As a result, the competition grew to be one of the biggest and most influential events where anyone can take part in with whatever topics they choose. In addition, it contributed to developing the concept of book design in an international book design forum. Since 2006, the General Administration of Press and Publication has given national government award to 10 books every three years. The aforementioned three awards are viewed as the major domestic recognition in the industry, providing a chance of competition and motivating young book designers to come up with creative works. Winners of those prizes occasionally show a good business report. They have been displaying outstanding talent in the community and took the status of the Chinese book design to global levels. Examples are Wu Yong, Lu Xiaoshang, Shiao Mage & Chen Zi, Jiao Qing, Zhu Yingchun, He Ming, Hone Wei, Yang Linqing, Ma Silui and Ren Jie & Pu Pan.

Though there has been the splendid advancement over the last several years, the Chinese book art is faced with many problems. More than 550 publishing firms and 10,000 magazine publishers are gushing out some 300,000 new and 100,000 reprinted publications. Overwhelming workloads force designers to deal with their jobs as they usually do. On average, they come up with 400 to 500 cover designs, so they have already lost motivations for creation and simply handle them. Some publishers just focus on changing book covers, driven by economic incentives. They just try to lower design costs, not caring about what is inside the cover. As a result, texts of such books are not interesting. Also, problems related to copyright, including illegal reproduction and copy, are widespread in the industry. Mass publications with low quality frequently fail to receive attention from the market and directly go to trashes. This vicious cycle puts countless book designers under the situation of excessive labor yet poor rewards, posing the risk of survival. We cannot sit still. Those conscious and responsible book designers always think about the meaning of making a book and do not neglect to reflect on themselves. They are building broader consensus in the industry, being ready to deal with difficulties. They all agree that they have to make concerted efforts to give books gorgeous outfit and a vital and interesting concept of the inner design. Designers should participate in conveying the message of texts and make textual information on pages have time-flexible meanings. This means that a book must give the poetic meaning of a text. Bottom lines of publishers should compromise the value of books, which will be materialized when creative works are passed down to the next generations. In a sense, design is an attitude. When a person who really cares design finds a book with meticulous design and abundant textual description, he or she wants to read it.

Now, the world has entered the era of Big Data. China is experiencing high growth in Internet penetration. According to reports by a government agency, the number of people going to the Internet was about 620 thousand in 1997, before reaching 684 million in 2015. Those who have access to the Internet on mobile devices were 620 million in the same year. It is 1,100 folds in 18 years. The surge of electronic media that are giving a shock to the traditional publication industry is a trend we cannot avoid. Most publishers are making efforts and putting financial resources into the new tool. A new way of reading brought by the novel technology of the 21st century will be prevailing and can be a desirable resonance to books. But there is an interesting fact that catches my eye: the traditional way of reading a book is gaining attention among young generations. They are reading well-known books. Those who value culture establish publishing companies and editorial planning firms, or collect money and cooperate with existing publishers to publish market-oriented, good-to-read books. These cases offer designers the chance to do away with publisher's stereotyped thinking and to work on their creativity. They are increasing the possibility of making a good book. Related examples are becoming more common.

Winners of "most beautiful book" are mostly freelance designers. The ratio between free lancers and those employed by national publishing companies is 10:3. Those who work freely have a bigger creation space compared with those hired. This clearly shows how a different working condition can make a difference to the results. There is another trend that we have to take note. National bookstores that do not respect "the bookstore culture" are decreasing, while such private bookshops that have a creative and vital book culture increasing. Young readers frequently go to and enjoy the second, not the first. Exemplary shops are All Sages Bookstore (万圣书店), One Way Street (单向空间书店) and Kubrick(库布里克书店) in Beijing; Librairie Avant-Garde (先锋书店) in Nanjing; The Mix-Place in Hengshan,(衡山和集书店) Shanghai; Fang Suo Commune (方所书店) in Guangzhou, Chengdu and Chongqing. Bookstore The Eslite Bookstore (诚品书店) of Taiwan is already operating in Suzhou and plans to open shops on the 62nd and the 63rd floors and B1 of the tallest building in Puxi, Shanghai. Art, reading and living are new topics for bookshops, these days. Although more and more people enjoy virtual reading, the demand on and popularity of paper-based books are growing. Readers are a big fans of books that are manually published by one-person publishers, or are limited editions. Recently, publications coming from sub-branded publishing companies. Books coming from Banana Fish Books (香蕉鱼书店), Pause Bread Press (加餐面包), Yan Shu book (罨书), L-A-B Design (连和部设计), and Youya Studio(友雅工作室) were interesting enough to bring in decent sales. University students design competition of China has been held four times by the cooperation between Book Design Art Commission of the Association of the Chinese Publishers and Graphic Design Art Committee of China Artists Association. The collaboration also made efforts for art schools to put book design in their curriculum. Book Design, a magazine published by Book Design Art Committee, optimizes academic research and plays as a good exchange platform for disseminating aesthetics of and world-leading ideas with regard to books. The Diamond Leaves(钻石之叶) book fair planned by Xu Bing(徐冰) of the Central Academy of Fine Arts, received very good reviews, becoming an exemplary case that showed public's interest in art books. Paperlogue, my book design study group, proposed the comeback to traditional reading and a new book-making movement.

Book art in China a long river of history that is experiencing new developments. Chinese designers have to show their respect to heritages left by their ancestors. Humbly and cold-heartedly, they should understand their progressive book-making attitude and inherit it with due respect, before passing it down to the next generations. At the same time, they must be open-minded so that they could learn from other cultures in different countries. They have to respect the way a development is made. Only when all of these are organized well, they can follow progressive flows. This also can be a drive that advances the Chinese book art. Tradition is neither a fixed way of copying nor an arrogant slogan. Generally, each ethnic design does not much deviate from the ground of its traditional culture. World-class designers are finding ways of expanding their culture under the modern linguistic environment. More than four decades of reforms and opening-up, book designers have materialized their idea of making books with passion, efforts and wisdom, by coming together – as if a rainbow with diverse colors show a wonderful harmony.

Let's create the beauty of a book and leave the meaning of "reading."

◆ 在雅昌人敬人书籍艺术工房讨论工艺

Collect the Precious

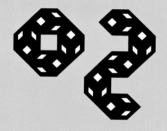

聚珍汇集

杉浦康平先生指出：『万事万物都有主语，森罗万象如过江之鲫，是一个喧闹的世界。一个事物与另一个事物彼此重叠层累，盘根错节，互为纽结，连成一张网。它们每一个都有主语，经过轮回转生与其他事物和谐共生，即共通精神气韵。』每次想到杉浦先生为银花杂志做设计，历经三十五年，就惊叹不已。

这是一个系统工程，每一册都有反映观察自然、生活、民俗、人类、社会的视角，汲取繁复纷杂的资讯。洋洋数百册，每册都可以成为一个独立的个体，却又如此统一于多主语的宏大的『银花』世界。

这些年我也设计了一批套书、杂志、丛书，从中国美术全集中国民间美术全集 敦煌石窟全集 到 高教设计专业创新系列教材 经济科学译丛 等，规模不算小，但远远达不到杉浦先生的气度，这里并不只是需要一种设计手法，更要有聚珍汇集、整合智慧的思维方法，我还在学习当中。

Kohei Sugiura said, "All things in this world have their subject. They are making the world bustling, like a fish crossing a river disturbs it. One thing is linked to another, forming a web among them. In summary, everything is with a subject, and its circulation builds a symbiotic relationship, thus making a sense of common atmosphere." I cannot be more surprised by the fact that my teacher has been working on the cover design of magazine Eunwha for 35 years. His design is a systemic process that collects valuable elements and turns them into one. Each of his cover designs reflects the perspective that views natural, living, ethnic, human and social factors, and absorbs complicating information. As such, hundreds of books published during the period can be regarded as an individual case, but, at the same time, be unified to a world of Eunwha.

Also, I have been making designs for complete works and a collection of books. I designed various books, including the Grand Collection of China's Fine Arts, the Grand Collection of China's Folk Fine Arts, the Grand Collection of Dunhuang, Design Textbook for Art Schools, Principle Economics, but cannot be a match to my master. His spirit and broad-mindedness are not within my reach.

I am still learning design skills and ways of collecting wisdom from him.

022

S ⇨ 260 × 340 mm
D ⇨ 1992-1994

中国民间美术全集
山东教育出版社＋山东友谊书社

《中国民间美术全集》设计于20世纪90年代初,恰逢中国跨入平版印刷(胶印)时代,这是我从日本学习归来的第一次设计,采用网格设计和编辑设计手法,与学者、编辑、印刷单位共同商榷互动,实现了许多设计愿望。封面和书脊建立统一的格式系统,以繁复的图像元素构成丰富多彩的民间艺术长廊。尽管纸张、工艺都很平常,但为充分还原图像四色印刷,让边缘线离切口5mm等,对印制提出了严格要求,在当时经历了艰难的沟通。1995年,此书获得第4届全国书籍装帧艺术展览整体设计金奖。

The Grand Collection of Chinese Folk Arts

Shandong Education Press + Shandong Friendship Bookstore

The book was designed in the early 1990s, when China entered the era of offset printing. Returning from Japan, he worked with scholars, editors and publishers, while trying diverse designs. A unified grid system was used in the cover and the back. Complicating figures formed a long alley of folk arts. Kinds of paper and crafts employed in the book were not special. The requests, including high quality levels of four colors and challenging sizes of empty space of 5mm in printing, made communications with printing staffers difficult. Those efforts were rewarded with Gold Award at the 4th National Book Design Exhibition in 1995.

S ⇨ 285 × 210 mm
D ⇨ 1998

中国现代陶瓷艺术

江西美术出版社 等

此套书是《中国现代美术全集·陶瓷卷》的重新装帧。设计更讲究古朴、典雅的个人化追求。书盒、包封均用特质纸，函盒书脊把陶艺家高振宇的青瓷瓶切割成五等份，以取得检索方便及趣味化的设计效果。包封以返璞归真的自然色白纸为基调，置入简洁的视觉图像，保留较大的空白，淡化多余的部分。书脊、内封冠以封面陶瓷器皿的概括归纳图形，形成本书各卷的识别标志，标志也渗透于文章、扉页、文字页、隔页与版权页之中，形成连贯统一的套书印象符码。

Contemporary Ceramic Art of China

Jiangxi Fine Arts Press and others

The book is a new design of the Ceramics of the Grand Collection of Contemporary Fine Arts of China, and its design pursues unsophisticated yet elegant personality. Book cases and covers use a specially treated paper, and each back of the cases carries each image of a cyanide ceramic work of ceramic artist Cao Zhenwu that is divided into five. Through this, he provided a reader with good design effects and a convenience when deciding which volume to choose. The book cover offers a simple visual image of white that is as natural as possible and shows a relatively wide span of empty space. Images of the ceramic ware are printed on backs and inner covers of the book. In particular, they are aligned according to their cover, so readers can easily identify which one is which. This system also applies to texts, a title page, main pages, inside page and publication right page, giving a feeling of being unified.

S ⇨ 215 × 285 mm
D ⇨ 2006

中国美术全集
人民美术出版社 等

**The Grand Collection
of China's Fine Arts**

----

*People's Fine Arts Press
and others*

025　S ⇨ 285 × 215 mm　D ⇨ 1999-2004

敦煌石窟全集

商务印书馆

此套书是为商务印书馆（香港）有限公司设计的。编辑要求体现出敦煌石窟的特征，并准确传递给读者敦煌的艺术韵味。请专家题写书名，以石窟唐草纹局部演变而成的标志贯穿全书。文字版赋色与白底黑线布局的图版做视觉分割处理。

The Complete Collection of Dunhuang Grottoes

The Commercial Press

This book, published by Hong Kong Commercial Press, was edited in such way that it describes the features of *Dunhuangshiku*, a historic earth cave, and distributes its artistic beauty in an accurate way. A calligraphic expert wrote the title on the cover of *Dunhuangshiku*, and all collections of the book were consolidated through symbols which are original patterns of the cave. Colors that appear on characters are visually separated from black-line images on the white background.

S ⇨ 290 × 420 mm
D ⇨ 2011

最后的皇朝——故宫珍藏世纪旧影

紫禁城出版公司

《最后的皇朝——故宫珍藏世纪旧影》七册一函,除总目外分别以宫殿、陵苑、帝后、宫廷、工业、军务等六册分卷收纳了千余幅珍贵的清末宫廷老照片。全套书整体采用中式筒子页包背装,内页纸张选用55g进口嵩高纸。这样的装订形式在增加翻阅手感柔软度的同时又不失厚重感。由于部分照片曾采用多次曝光的技法,画面灰色层次细腻丰富,且年代久远的照片会不同程度地产生斑驳的附加肌理,因此双色印刷手段能很好地呈现出老照片的原貌。版式在设计上也借用双色技术,有层次地梳理出中、英、日三种语言的图版资讯;并在文章前版面上点缀专红色,以求尽可能体现出与照片年代相符的古籍气质。本书隔页不仅包含细分类别的提要内容,而且其筒子页背面是反印的具有特色的本章节照片,翻动时透过轻薄的纸张分割章节的图像隐约可见。每一卷书的目录前还选用了半透明特种纸张,用专金色来印制一幅照片的负片效果,读者可从不同的角度交替感受照片成像的前后两种状态。

The Last Dynasty -The Palace Museum Collection of Century Old Photos

----

The Forbidden City Publishing House

There are seven volumes in this series, each of which is about: table of contents; palaces; royal tombs; feudal lords; royal court; manufacturing; and, military affairs. More than 1,000 rare images taken at royal court of the Qing Dynasty during its late years are carried in the book. The full set utilized the Chinese *book binding* style, and the inner pages used imported natural paper (55g). This type of binding gives readers a smooth feeling to their hands when they turn the pages, while keeping a sense of weight. Some pictures in the book are in multi-exposures, so their brightness is rich and outstanding. Old images are added with different variegation textures, and two-color printing method was used to exhibit their original feeling. Editorial design also used the two-color technology, carrying images' information in Chinese, English and Japanese language. In addition, texts are marked with red stars. This is to create a mood that goes with old books, considering specific periods of time in each image. In the writing paper of the book are detailed explanations by type. On the backs of double-folded pages

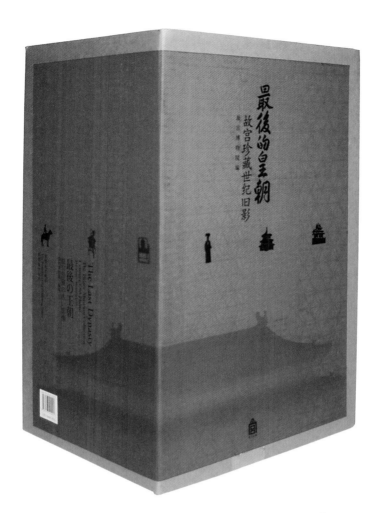

are reversely printed with unique images, so that readers can dimly see pictures that divide chapters and clauses through thin and light paper, when turning pages. Before the table of contents of all volumes of the book is a special translucent paper with gold stars on it. This was done for a negative print effect. Thanks to it, readers can read the book, while changing the viewing angle in front of or behind an image. Inspired by diversified and complicating geometric window patterns used in the construction of royal court of the dynasty, seven new patterns were designed. Each of them was used for book cover and inner pages of each designated copy, and simple and grave color tone were added on the silk-texture material

借助清代宫廷建筑中复杂多变的几何窗饰结构灵感,设计者为该套书重新设计了七款纹样,分别应用在七册书的封面和内页设计中,辅之以单纯厚重的色彩后印刷于丝绢质装帧材料上,精致的纹样与有光泽的材质相互协调。打开环衬后,则为与封面纯度色相一致的高染色宣纸扉页,为进入正文增加了一层过渡关系。封面的螭纹与提取的本册影像元素既能区分七卷内容,同时又有整体的系列感。为每本书所配置的一个函夹,不仅方便从总书函中抽取,还可以在桌面摊开为翻阅提供保护。由于除了总目的其余六卷没有先后顺序,因此这七款函夹的书脊图案以各种顺序插放在外函套中,均可连贯一体。把六册分卷的几何纹样通过金色烫的技艺附着在外函套红、金双色装帧面料上,这六组华丽的圆形纹样中间浑厚有力的书名题字取自康、雍、乾三代帝王的书法作品。庞大体量的内容通过七卷书汇集成一函后,红、金、灰三色浑然一体的视觉感受与"皇朝"主题相契合。

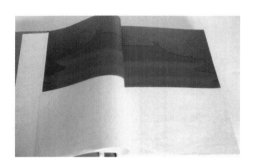

of book binding. All of these are designed for sophisticated patterns to go well with glossy materials.

When you turn the first page, you will see a title page, made of rice paper, in the same color with the book cover. The page is designed to give a feeling of a door leading to the main body. Dragon patterns of book covers and elements drawn from images in the book are points by which you can tell one from others, and, at the same time, become a common ground that makes you feel that they are a different volume of the series book. A stand put in every set helps you to pull the book out and put it on a table, while protecting it. Six copies of the collection, except the volume of the table of contents, were not serially designed, so patterns on their back form a good picture regardless of their location among them. Geometric gilt patterns of the six volumes are inscribed on red and gold-color fabric. In the middle of gorgeous six circular patterns are book titles, and the font of them are hinted from calligraphic works of three major emperors of the dynasty, Kang Xi, Yong Zheng and Gan Long. This much amount of contents are well organized in the seven publications of the book, so that three colors of red, gold and gray become united to give an excellent visual impact and to display the theme of the imperial court.

高教设计专业创新系列教材

高等教育出版社

A Creative Series of Textbooks for Art and Design Colleges

Higher Education Press

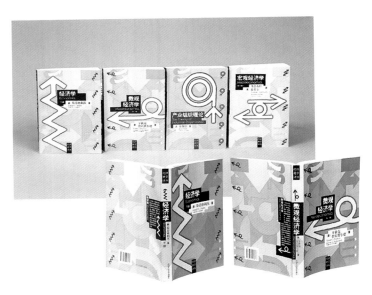

028

S ⇨ 185 × 260 mm
D ⇨ 1997-2015

经济科学译丛

中国人民大学出版社

Collection of Translations of Economics

People's University Press

029

S ⇨ 185 × 260 mm
D ⇨ 1997-2015

熊十力全集

湖北教育出版社

Collection of Xiong Shili

Hubei Education Press

030

S ⇨ 180 × 260 mm
D ⇨ 2009-2016

民国文献资料丛编

北京图书馆出版社

Collection of Materials in the ROC Period

Beijing Library Press

031

S ⇨ 210 × 285 mm
D ⇨ 2005

周原出土青铜器

四川出版集团+巴蜀书社

Ancient Chinese Bronze
Vessel of Zhouyuan

Sichuan Publishing
Group + Bashu Press

032

S ⇨ 145 × 210 mm
D ⇨ 1998

中华文化通志

上海人民出版社

Collection of Traditional
Chinese Culture

Shanghai People Press

033

S ⇨ 210 × 285 mm
D ⇨ 2005

陕北出土青铜器

四川出版集团+巴蜀书社

Ancient Chinese Bronze
Vessel of Shanbei

Sichuan Publishing
Group + Bashu Press

034

S ⇨ 270 × 370 mm
D ⇨ 1996

20世纪下半叶中国书法家全集

四川出版集团+巴蜀书社

20th Century Chinese
Contemporary Cal-
ligraphic Art

Sichuan Publishing
Group + Bashu Press

035

S ⇨ 220 × 285 mm
D ⇨ 2007

钱学森书信

国防工业出版社

Letters of Qian Xuesen

National Defense Indus-
try Press

Entry into Book Binding / 1978−1989

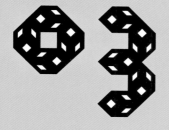

装帧入门

「文化大革命」结束后的一九七八年，我进入北京的中国青年出版社，开始了与出版相关的工作生涯。当时我的工作以创作文学插图为主，为真实反映原著的历史风貌、人文特征，经常下乡采风，收集素材，受益匪浅。以后逐渐进入封面图的创作，即所谓的装帧。那个年代，中国大量的印刷品是依赖于铅字排版、凸版印刷。每次印封面时，我都要跟着师傅在机台调油墨看印样。设计就是用水粉或水彩画封面，文字全靠手写。那时经济困难，只画封面，没有封底，只准用三色，依靠相互压印呈现第四色或第五色，有点像印版画，很有意思。条件虽然差，但学到了很多东西。尤其从许多前辈同行和印厂师傅那里学到了很多装印方面的知识和一丝不苟的工作态度，受用至今。回想做这些品质并不高的封面的过程，还是挺令人享受的。

I started my career at China Youth Press in 1978. My jobs in the publishing industry were mainly about illustration in literature works. I tried to find inspiration and materials in daily life to vividly describe the features of historic figures and personalities of characters. Those efforts made during the time have remained as quite a good amount of nutrition since I became a book designer. Then, I began to create cover images, which was called book binding. Back then, the Chinese printing industry was largely dependent on the lead-type printing, so whenever I was printing a book cover, I had to follow a printer operator when deciding on the amount of ink for a proper sample. The design was mostly poster-color, or watercolor paintings. Characters were made manually, 100%. Economic situations were poor, so I came up with only Table 1. I did not work on Table 4. Only when I carried out estampe, a French meaning a type of print, I could have four or five different colors. This was very interesting to me, as it looked like an engraving. Though the working environments were very unfriendly, I learned many lessons from those experiences. Particularly, I came by much information about book binding and ways of meticulously doing works from those who had more knowledge than I had as well as those who handled printing machine. The learning still works for me. In hindsight, the experience working under an extremely inadequate is worth trying.

036 S ⇨ 184 × 130 mm
D ⇨ 1979

卓娅和舒拉的故事

中国青年出版社

Old Story of Zhuo Ya and Shu La

China Youth Press

037

S ⇨ 184 × 130 mm
D ⇨ 1979

五个小罗汉（与刘宇廉合作）
外文出版社

Five Little "Arhats"

Foreign Languages press

Today they have learned that the physical culture teacher is not around for the school's martial arts team practice. They decide to make a surprise attack on th team. Why not? The team shows no intention of admitting the "arhats."

Seeing that the old door k is dozing off, they steal int school grounds. Members o martial arts team are practisi the small campus shaded by trees.

The leader's name is Wei Wei. He often goes to Jingshan Park to watch people practice martial arts, and can imitate some of their movements. Who would dare to offend him by doubting his skill? The four others treat him as the boss.

Look! Here they come. They [ab]out the same height and size. [They] strut about and put on grand [airs.] Ha! The most eye-catching [thing] is their five shaven heads — [shi]ning in the August morning [sunsh]ine.

038

S ⇨ 183 × 130 mm
D ⇨ 1983

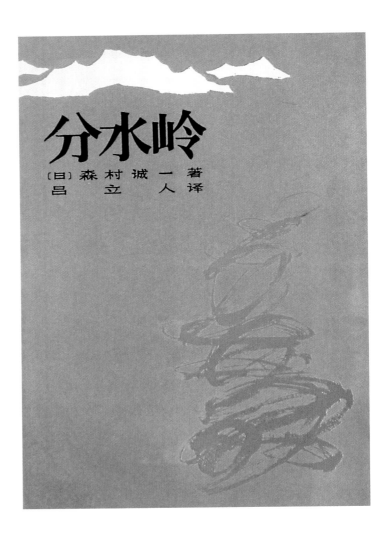

分水岭
宝文堂出版社

Watershed

Bao Wen Tang Press

039   S ⇨ 130 × 183 mm
D ⇨ 1983

括苍山恩仇记
中国青年出版社

Revenge on Mt. Kuo-cang

China Youth Press

040　　　D ⇨ 1980　　　采桑图

**Selected Poems:
Plucking Mulberry
Leaves**

Illustration

D ⇨ 1980

细草微风岸，
危墙独夜舟

Selected Poems:
A Gentle Breeze,
A Lonely Night

Illustration

041   S ⇨ 203 × 140 mm
       D ⇨ 1984

生与死
中国青年出版社

Life and Death

China Youth Press

042

S ⇨ 185 × 206 mm
D ⇨ 1986-1992

邮票中的世界名人

中国青年出版社

World Famous People on Stamps

China Youth Press

043

S ⇨ 185 × 206 mm
D ⇨ 1986-1992

邮票中的世界名画

中国青年出版社

World Famous Paintings on Stamps

China Youth Press

044

S ⇨ 150 × 203 mm
D ⇨ 1985

"亚细亚"之恋

中国青年出版社

Love with Asia

China Youth Press

045

S ⇨ 290 × 420 mm
D ⇨ 1980

蛇类

科学出版社

Snakes

Science Press

046

S ⇨ 150 × 203 mm
D ⇨ 1983

不倒的红旗

中国青年出版社

The Red Banner Forever

China Youth Press

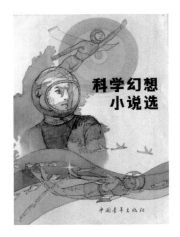

047

S ⇨ 140 × 203 mm
D ⇨ 1983

科学幻想小说选

中国青年出版社

Album of Science Fiction Novel

China Youth Press

048

S ⇨ 140 × 203 mm
D ⇨ 1979

播火记

中国青年出版社

Seed Sower

China Youth Press

049

S ⇨ 140 × 203 mm
D ⇨ 1983

烽烟图

中国青年出版社

The Billowing Beacon

China Youth Press

050

S ⇨ 130 × 184 mm
D ⇨ 1983

王安忆中短篇小说集

中国青年出版社

Wang Anyi's Novellas and Short Stories

China Youth Press

051

S ⇨ 130 × 183 mm
D ⇨ 1980

强者

中国青年出版社

Powerful Man

China Youth Press

052

S ⇨ 140 × 230 mm
D ⇨ 1986

无悔的追求

中国青年出版社

Regretless Chasing

China Youth Press

053
S ⇨ 140 × 203 mm
D ⇨ 1985

单身汉的乐趣

中国青年出版社

The Happy Bachelor

China Youth Press

054
S ⇨ 140 × 203 mm
D ⇨ 1984

无头骑士

中国青年出版社

A Knight Without Head

China Youth Press

055
S ⇨ 140 × 203 mm
D ⇨ 1985

猎人的姑娘

人民文学出版社

The Hunter's daughter

People's Literature Press

056
S ⇨ 130 × 183 mm
D ⇨ 1980

魂兮归来

中国青年出版社

Summon Spirit

China Youth Press

057
S ⇨ 128 × 183 mm
D ⇨ 1987

黑色飞机的坠落

宝文堂出版社

Black Jet Down

Bao Wen Tang Press

Seek Learning and Find a Way / 1989–1998

求学寻道

『装帧』是时代的产物。二十世纪八九十年代，出版商品化，把书衣打扮当成利益最大化诉求的装帧定位，弱化了文本阅读功能的编辑设计力量的投入。我苦恼于这种装帧观念与国外先进出版物之间的差距，并觉得这是中国书籍艺术跨入新阅读时代的意识阻隔，于是我增强了求学的渴望。二十八年前赴日，我有幸到杉浦康平老师的设计事务所学习，他改变了我对书的设计观念。他使我顿悟，优秀的书籍设计师不仅仅满足于书的外在，而且应在文本的篇章节句中寻找书籍语言表演的空间场所和叙述故事的时间过程，让视觉资讯游走迂回于页面之中，让书纸五感余音缭绕于翻阅之间……感染读者的情绪，影响阅读的心境，传递着善意设计的创造力。他一再强调一本书不是停滞在某一凝固时间的静止物体，而是为读者构造和指引诗意阅读，富有意境的生命体，我如醍醐灌顶，反省初衷。

Book binding is a product of an era. As publication became an area of business from 1980 to 1990, it was purported to decorate books beautifully for profits. Accordingly, the drive to help readers read texts weakened. I was concerned about the contradictory situation in the Chinese industry, which was a real problem in China compared with advanced nations. During this period, the big difference became an ideological hurdle for the Chinese book art to enter a new era of reading. My thirst for learning went greater and greater. To quench the thirst, I finally decided to go to Japan, where I luckily picked up an opportunity to learn from Kohei Sugiura. That is the story 28 years ago. The master fundamentally changed my perception of book design. He told me that a respected book designer should not only decorate the outside of a book but also find a time of story and the space expressed by the language in between sentences as well as phrases. He had me understand that I have to fulfill my work in such ways that make visual information on pages go alive; page-turning sound vivid, so that it can touch the emotion of a reader; reading ensures a good influence; and, a good design creativity transfers to others. The gist of his theory is, "Books are not a thing in a halted time. We have to design books for readers. Book design should be done in ways of ensuring poetic reading, ultimately being a living creature on an artistic level." His idea made me reflect on myself and go back to the original attitude.

058

S ⇨ 230 × 150 mm
D ⇨ 1990

讲谈社 | 家

Home

Gangdamsa (Japan)

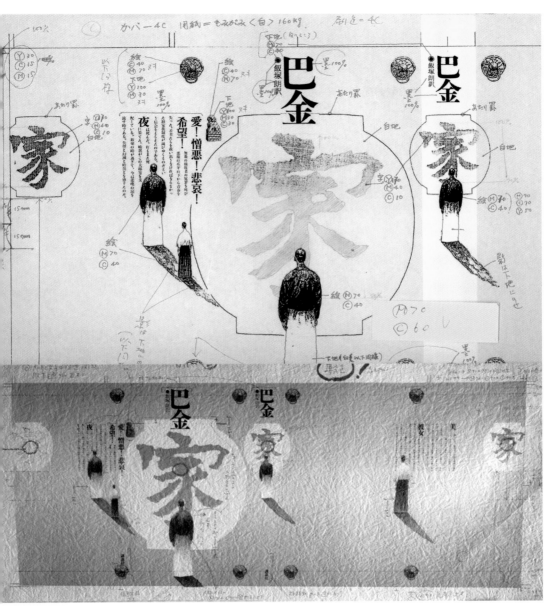
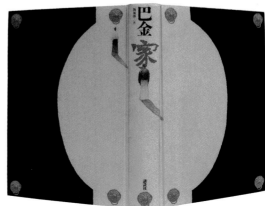

059   S ⇨ 185 × 130 mm
                 D ⇨ 1992

我是猫 | 中国少年儿童出版社

I Am a Cat

China Children Press

S ⇨ 203 × 140 mm
D ⇨ 1995

赤彤丹朱
人民文学出版社

本书在封面设计上没有运用具体图像，而是以略带拙味的老宋体书名文字围绕排布成窗形，字间空档用银灰色衬出一轮红日，显得遥远而凄艳。它无言地召唤着读者的关注，继而引领读者将设计家专门请作家撰写的四句提示语细读一番，最终让人欲罢不能。视线差的渗透取决于视觉形态的暗示性。这种意义上的表现力，往往通过视觉形式感来传达某些难以言表的意味、情绪或气氛。它具有抽象品格，特别讲究视觉形态的隐喻性。

Different Kinds of Red

People's Literature Press

The book has no image on its cover. Instead, it puts a shape of window there, with the title in the Ming style font on the side. The empty space between characters is in silver gray to express mimil a glowing sun, giving a sense of beautiful yet pathetic. This book drew attention of readers, quietly. The designer of the book had the author write four tips, which glued readers' eyes to it. The flow of eyesight comes from the implication in visual format.

This type of expression often conveys a taste, an emotion and a mood that cannot be visually described. It bears an abstract dignity and a metaphor in the visual format.

Black and White

China Youth Press

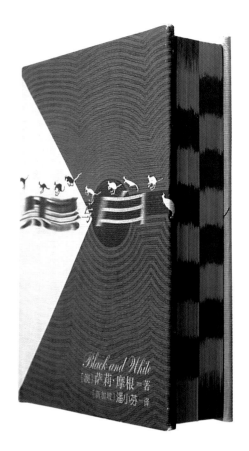

S ⇨ 205 × 105 mm
D ⇨ 1996

书籍设计四人说 | 中国青年出版社

本书由吕敬人和宁成春、朱虹、吴勇四位设计家共同编著、设计和制作。这本书将四个人的设计作品融进四个人关于书籍设计理解的对谈录。编辑思想灌入书籍设计新观念即"资讯的再设计"。本书用 25 种纸张印制而成，是一本强调书籍形态和功能相结合的书，也是真正体现从装帧到书籍设计这一概念改变和实践过程的书。

LNWZ: Four People Talking About Book Design

China Youth Press

This book is collectively edited, designed and made by Lu Jingren, Ning Chengqun, Zhu Hong and Wu Yong. It is their talks about design and understanding of book design. It is "re-design of information", led by the combination of new conception of book design and editorial thoughts. It emphasizes the mixture of forms and functions of books, by using 25 different kinds of paper for printing and book-making. This book truly shows the change and practice of concepts from book binding to book design.

◆ 1—3
1996年书籍设计四人展现场
The exhibition scene of
LNWZ in 1996

影响我一生的两位恩师
贺友直 / 杉浦康平

Two Masters Who Influenced to My Life
He Youzhi / Kohei Sugiura

贺友直

1973年，我在北大荒幸运邂逅正在挨批的"资产阶级反动学术权威"贺老师。那时50岁的贺老师正值创作高峰期，他创作的连环画《山乡巨变》的影响可以说覆盖了整个美术界。"文化大革命"伊始，他的创作权利戛然而止，瞬间成了整日挨斗的"牛鬼蛇神"，接着被下放到农村接受劳动改造，多年后被派到北疆与工农兵三结合创作组。就这样，在我下放的农场天上突降下来一个我们学画时特别崇拜的贺"姥姥"（如同《红楼梦》中的宝玉那样，朝思暮想着天上掉下来一个林妹妹）让我们这帮绘画青年乐得屁颠屁颠的（手舞足蹈）。我与另两位知青有幸成为三结合创作组的成员，开始了与这位"黑帮分子"同吃、同住、同劳动、同创作的朝夕相处的日子。365个日日夜夜，我们在生产队体验生活、收集素材、研讨脚本、塑造人物、构想情节、制定手法、完成画稿，一年后，连环画《江畔朝阳》出版了，从此我们成了贺老师名副其实的学生。贺老师的艺术追求、细微的生活观察、严谨的创作方法，使我们这些没有经历过专业训练的绘画青年茅塞顿开。他勤于思辨、认真做事的态度；他疾恶如仇、端正做人的秉性，使他成了我们人生正道的指引者，几十年来他像我的父辈，亦是良师益友，更成了心灵沟通的莫逆之交。

30年前，我为一本小说画了一套插图，寄给贺老师请他提意见，不久收到回信，启封一看，我惊呆而感动：严整的笔迹，长长的点评，亲手绘制的范例，他把插图的要点、立意、方法论，深入浅出地娓娓道来，针对我的缺陷与不足给予了中肯的批评。我的书籍设计生涯就是这样在贺老师言传身教中熏陶觉悟过来的。自在北大荒认识贺老师后的40多年里，多少封书信，多少回授艺，多

◆ 在央视读书节目介绍贺老师的作品集《贺友直画三百六十行》
During an interview by CCTV, Lu introduced Old Occupations of Shanghai, an important works of He Youzhi, to the audience.

◆ 贺老师为我的插图创作指点迷津
He Youzhi tried to guide my painting creation through demonstration in his letters.

少次恳谈已经记不得了。弹指一挥间，进入古稀之年的我几经风风雨雨，坎坷与幸运，失落与收获，事业与生活的每一个关键点上都有恩师的点拨和指引，我无法忘却。

贺老师一生创作了影响几代美术家的众多佳作，《山乡巨变》《小二黑结婚》《朝阳沟》《李双双》《十五贯》《白光》……得奖无数，享誉世界。我有幸为他设计了《贺贺友直画三百六十行》《贺友直自说自画》《杂碎集》三部曲。他那独特的贺式视觉导演手法和与生俱来的人物表演才华，把彼时、彼地、彼情、彼景一五一十、活灵活现地描绘出来，也许一些研究社会史的学者只有从他的画面里才能寻见时空倒转的记忆。善于观察生活，喜欢琢磨事态的贺老师随时把摸社会的脉搏，透析人世间的美恶善丑，他的脑子从来没有停息过。我在书籍设计中应用的编辑设计和"书戏"理念都是受到他的影响，我今天取得的些许成果也离不开恩师的悉心指点。

He Youzhi

In 1973, I ran into He Youzhi in Beidahuang, who was denounced as "a scholar who advocated the proprietary class." He was then 50 and in the prime time of his creative activities. His picture book Shanxiangjubian was influential enough to affect the whole fine art community. But, the Cultural Revolution stopped him from producing creative works. He had been one of the best designers, but became a subject of all-day monitoring after the political event. He was soon sent to remote areas as he was sentenced to "labor reform." Several years later, he was transferred to Beijing to be under the censor by Chinese soldiers. Those serving the same sentence, including other young designers and me, were very happy to be with him. My two friends and I somehow made a small creativity group. We had meals, labored and carried out creative activities together, day and night. We acted as if we had been his assistant: collected materials; discussed scenarios; created characters; and, came up with stories. We also drafted production process and completed paintings and scripts. One year later, we printed a picture book Kanpanjoyang. This accomplishment led us to think that we were truly his students.

His pursuit of arts, meticulous scrutiny in everyday life, and rigid and diligent ways of creation were like a new logic, or a huge wave to young painters who did not attend professional training courses. He showed earnest attitudes to his works; did not stop thinking of his jobs; did not look the other way, when he saw unjust things. He became a beacon to us. The attitude to life that I learned from him back then is still alive. For the past several decades, he had been my master, friend and soul-mate, and sometimes even father.

Three decades ago, I happened to draw an illustration for a novel. I sent the book to Mr. He for his comment. Soon, he gave me his thought. I could not be more appreciated for his heart-felt letter. His writing was well organized, supported by logical backgrounds and examples. He suggested his idea, which was about ways of making a summary, planning and methodology. They were very productive as I could know my weakness and shortfall. It hit the mark. My book design career was advanced by his comments and behaviors. Since I met him in Beidahuang 40 years ago, we have exchanged countless letters and discussions. Looking back on my past 70-years, I met many ups-and-downs, including hardship and luck, disappointment and happiness. His advice and consultation have been a truly big boost to me.

He created great works that would be influential to next generations of artists. Ever-lasting passion for creation led to a number of awards and a big reputation in the world. His textbooks will be time-honored materials, including *Shanxiangjubian, Xiaoerheijiehun, Zhaoyang Gou, Li Shuangshuang, Shiwi guan* and *Baiguang*. I designed *Old Occupations of Shanghai by He Youzhi, He Youzhi's Words and Paintings, He Youzhi's Collection of Essay* for him. His unique perspective and talent in describing characters well explained diversified time, places, events and landscapes. Any scholar in the field of social history would find his or her memory going beyond time and space in the screen he staged. While sharply observing everyday life and enjoying situational analysis, he occasionally grasped the context of a society and tried to analyze all things in the human's world. He never stopped thinking. I can even say that there was nothing that was not impacted by him in my editorial designs and "entertaining thought about books." I owe him my humble success as I see it today.

贺友直

浙江宁波北仑人，中国著名连环画画家、线描大师。他自学绘画，自 1949 年起开始画连环画，从事连环画创作 50 多年，共创作了百余部连环画作品，对中国的连环画创作和线描艺术做出了重大贡献。《火车上的战斗》在 1957 年全国青年美术作品展览中获一等奖。《山乡巨变》在 1963 年文化部与中国美术家协会举办的第 1 届全国连环画评奖中获一等奖，被誉为中国连环画史上里程碑式的杰作。代表作还有《白光》《十五贯》《小二黑结婚》等。出版有《贺友直谈连环画创作》《杂碎集》等专著。

He youzhi

Coming from Ningbo, Zhejiang province, he was a famous Chinese painter of picture books and master of line drawing. Self-taught, he had begun drawing pictures since 1949. For more than five decades, he made a significant contribution to picture book making and line drawing in China. His work A battle on the Train won the first prize of National Artworks by Young Artist in 1957; Shanxiangjubian won the first prize in the first National Comics Contest, co-organized by the Cultural Ministry of China and China Artists Association in 1963. Particularly, this work was acclaimed as a milestone book in the Chinese comic book history. His representative books are Baiguang, Shiwu Guan, and Xiaoerheijiehun. He also published *He Youzhi's Collection of Essay* and *Creating a Picture Book by He Youzhi*.

063   S ⇨ 178 × 276 mm
      D ⇨ 2004

**Old Trade of Shanghai
by He Youzhi**

*Shanghai People's Fine
Arts Publish House*

贺友直画三百六十行

上海人民美术出版社

eek learning, and find a way 1989-1998

◆ 和杉浦康平、安尚秀在敬人设计工作室合影
with Kohei Sugiura and Ahn Sang-soo in Jingren Art Design Studio.

杉浦康平

1989年，我有幸在日本杉浦康平先生的事务所学习，从这一刻起我才慢慢领悟书籍设计的真正概念和其中含有的知识容量。那时我在国内已经担任过十多年的装帧工作，当时的装帧不能全方位介入书籍的整体运作中，更谈不上触及文本的视觉设计——这是属于文字编辑的不可侵犯的领地。

在杉浦先生的设计过程中，书籍设计不是简单的装饰，我因此有了新的深切体悟。他让我明白所谓书的设计均是经过设计者与著作者、出版人、编辑、插画家、字体专家、印制者不断讨论、切磋、修正中产生的整体规划过程，尤其是杉浦先生对文本的解读，都有他独到的见解，更是以自己的视点与著作者探讨；再以编辑设计的思路构建全书的结构；以视觉资讯传达的特殊性去弥补文字的不足；以读者的立场去完善文本传达的有效性；以书籍艺术性的审美追求，着重于细节处理和工艺环节的控制；以理性的逻辑思维和感性的艺术创造力把书籍的所有参与者整合起来，发挥出各自的能量，汇集大家的智慧和一丝不苟的态度来做一本尽善尽美的书。我在国内从未体验过这样的做书经历，书籍设计师的这种专业性令我惊讶，更是引发我竭尽全力去关注，并参与一些书籍设计的全过程。这也使我重新认识和定位自己，重新界定设计师做书的目的性和责任范围，认识书籍的装帧、编排设计、编辑设计三位一体的设计理念之重要。设计者在其背后的知识的铺垫、视野的拓展、理念的支撑也十分重要，杉浦先生让我开始明白：作为书籍设计师除了提高自身的专业素养外，还要努力涉足其他艺术门类的学习，如目能所见的空间表现的造型艺术（建筑、雕塑、绘画），耳能所闻的时间表现的音调艺术（音乐、诗歌）；同时感受在空间与时间中表现的拟态艺术（舞蹈、戏剧、电影）。他引领我走进书籍设计之门。

杉浦先生说，评价一个设计师是否优秀，不仅要看他是不是有能容知识的"大坛子"，而且要看他在需要的时候能不能随时拿出来——是经过去伪存真，去粗取精，经过消化的东西，并使其成为创意的智慧点，能赋予个性的东西，这是一种跳跃性的思维。设计不仅仅是对技巧物化的高低评判，更是设计之外的知识的展示和修养的显露，犹如绘画界的一句俗语"功夫在画外"的道理。杉浦先生引领我走出"设计"从而获得了更为开阔的设计天地。这种出自学习建筑设计喜欢追根寻底的习惯，使他具有严谨的逻辑思维能力和对资料精度的严格要求。来自自然体验的噪音学说和东方的混沌思想，加上西方科学逻辑思维的严密性，形成了杉浦设计公式：艺术×工学＝设计^2。

杉浦康平先生学识渊博，思维敏捷，他的兴趣广泛，视野高远。他专一好学，没有停息过

片刻。他总是在往前走,虽然人生已经度过八十载,却仍像20岁的年轻人一样渴望新知。他从不求名利,不兴空谈,鄙视权力。他扎扎实实做学问,实实在在做设计,拥有令人敬佩的做事做人的态度。从20世纪50年代至今,他所创作的作品不计其数,成就斐然,独成流派,在书籍设计方面影响着日本、亚洲乃至世界。他凭借着对东方文化的酷爱,跨出自身国界疆域的局限,用放眼汉字文化的视野,彰显他的影响力——他那源自深远东方并超越国界的文化感染力,还有他的人格魅力。他在亚洲各国有无数仰慕者。

真正的艺术家从不会自命不凡,但真正做到这一点的人并不多。杉浦先生的艺术天地就是他生命中的全部,不求取任何虚空的名誉地位,不加入任何国际国内小团体,默默在艺术天地耕耘60余年。近30年来,我每时每刻受教于杉浦康平先生,他的学识、人品、专业态度、治学精神使我受益匪浅,他作为东方文人的傲骨令我钦佩,能做他的学生是我一生的幸运。

Kohei Sugiura

In 1989, I found an opportunity to learn from Kohei Sugiura at his workshop in Japan. Since then, I could gradually understand the real concept of book design and the knowledge staying inside it. By the time, I had already experienced book binding in China for over 10 years. The book-making process was not equal to the overall book planning back then. It was less likely to change the perspective design of texts. In addition, text editing is just untouchable.

His book design was not a simple decoration, which made me experience new and very special events. He told me that book design consisted of all relevant processes. He added that it was an overall planning stage where designers, authors, publishers, editorial illustrators and typo-experts and printing workers could ceaselessly talk, negotiate and modify. In particular, he had his own idea about interpreting texts and talked with authors about his opinion and analysis. He complemented gaps in texts, by constituting a book with an idea of editorial design and utilizing the peculiarity of conveying visual information. He made efforts to raise the efficiency of text delivery and pursued aesthetic appreciation of book's artistic value. Delicate treatment for detailed parts and technology control were also important. He collected abilities of all participants in a book through reasonable thinking and sensitive art creativity, so that they could fully partake in works. He also made their wisdom lead them. He always tried to do his best in all processes to come up with a perfect book, relentlessly pursuing virtue and beauty. I had never seen such an attitude in China. This level of his professionalism not only surprised me but appealed to me the attractiveness of book design more deeply. Since then, I took part in all processes of

064

S ⇨ 210 × 285 mm
D ⇨ 2006

疾风迅雷 —— 杉浦康平杂志设计的半个世纪

三联书店

Wind and Lightning - a Half - Century of Magazine Design by Sugiura Kohei
----
SDX Joint Publishing

065

S ⇨ 210 × 285 mm
D ⇨ 1999

杉浦康平的设计世界——
注入生命的设计

河北教育出版社

Design Works of Sugiura Kohei

Hebei Education Press

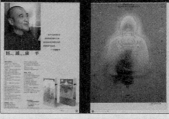
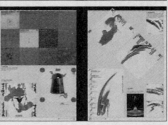

book design. Thanks to it, I understood book design again and find my location in it. I was also able to identify the objective and the responsibility scope of book-making. I could know the meaning of uniting book binding, lay-out design and editorial design into one and grasped the importance of design thought. Designers should have diverse background knowledge, broad viewpoint and perception. He emphatically told me that a book designer has to develop his or her own talent, but also learn enthusiastically from other fields of art: formative arts (architect, sculpture and painting) that is visible and despicts space: other arts (music and poetry) that is audible and expresses time: and, mimesis arts (dance, play and movie) that can be felt within space and time.

Thus, he led me to a world of book design.

He argued, "Excellent designers have to have a big container that can hold diverse knowledge. They should be able to utilize what is inside, whenever necessary. The knowledge can help them to address current problems they face and make something raw sophisticated. Things that have undergone processing should come in creative wisdom." Design is evaluated not only by levels of applied skills but intellect and practice. People in the painting community say, "Real skills exist in the outside of a painting." I think there is a common ground between the two thoughts. He had me overcome "design" and acquire more spacious

"design world." The habit of finding a cause began forming with the study of architect design. This attitude brought high levels of perception on elaborate logical thinking and data accuracy to Kohei Sugiura. This was then added to the noise theory, the chaotic ideology in the East, and scientific, logical thinking of the West. It became a ground of his

066

S ⇨ 150 × 210 mm
D ⇨ 1999

造型的诞生

中国青年出版社

Birth of Fine Arts

China Youth Press

design equation "Art × Engineering = Design²."

Kohei Sugiura was an erudite, fast thinker, interested in broad areas and very wide perspectives. He likes to learn and never stops learning. Throughout his life, he has been marching forward. Now he is in his 80s, yet still hungry for new things, like those in their 20s are. He has never coveted fame or money; has never spoken in vain; and, despised submission to power. He diligently carried out his research; put focus on practical designs; and, retained proper ways of doing works and attitude. People paid respect to him for these. Ever since the 1950s, he released countless works and made great achievements. He established his own school. He is influential in Japan as well as other parts of Asia and the world. His distinguished interest in Eastern culture expanded his activity areas from Japan to other regions whose culture is based on Chinese characters. The impact of his academic ideas comes from his charming personality and the cultural influence that goes beyond the East and borders. This is why so many people all over Asia pay respect to him.

True artists never regard themselves as excellent. There are not many who want to be recognized as such among them. Kohei Sugiura did not pursue any reputation or coveted status. He did not join any domestic or international association or organization, though he was viewed as an outstanding figure. Over the last 60 years, he has trodden his own path. For almost 30 years, there was not a moment in which I was not influenced by him. His diverse assets (academic knowledge, personality, professional attitude and willingness to learn) enabled me to think about many things. It is just my luck to know such a wonderful person in my life.

067

S ⇨ 150 × 210 mm
D ⇨ 2016

多主语的亚洲——杉浦康平设计的语言

中国青年出版社

Sugiura Kohei's Words and Thoughts on Design

China Youth Press

Kohei Sugiura's formula

艺术 × 工学 = 设计²

Art *Technology* *Design²*

[艺术×工学＝设计²] 就是用感性与理性来构筑视觉传达载体的思维方式和实际运作规则。艺术，塑造精神的韵；工学，构筑形态的物，两者蕴含着内在与潜在的逻辑，形神兼备的设计可达到原构想定位的平方值、立方值，乃至n次方的增值结果。研究概念：

●Ⓐ **书籍设计 [Book Design]**：认知装帧、编排设计到编辑设计三位一体的Book Design的时代必要性。●Ⓑ **编辑设计 [Editorial Design]**：学习文本信息传播控制的逻辑思维和解构重组的方法论，完成文本再造过程，达到阅读与被阅读的最佳关系。●Ⓒ **信息视觉化设计 [Infographic Design]**：掌握信息视觉化传递的设计思维和视觉信息图表设计的信息整合和表现方法。●Ⓓ **手工装帧 [Book Binding]**：探讨纸质媒介特征，回归手工装帧的手段，认知书籍阅读与物化设计的关系。

The equation, Art × Engineering = Design², means building ways of thinking that deliver visual media through emotion and rationality and rules of actual operations. Both the rhythm that embodies arts and spirit and the materials that establish engineering are implicit and have potential logics. Through the design that carries forms and sprit, the value that has already been set can increase by square, cube, triple and quadruple. Research concepts are as follows;

●Ⓐ **Book Design:** It perceives that a book design is currently in need of integrating book binding, lay-out design and editorial. ●Ⓑ **Editorial Design:** It is a methodology that re-configures the logical thinking and the structure that control the delivery of textual information. It constructs most ideal relationship between reading and being read, by way of text re-configuration.
●Ⓒ **Infographic Design:** It grasps expression methods that integrate the visual infographic design information and design thinking that visually conveys information. ●Ⓓ **Book Binding:** It explores features of paper medium and perceives the relationship between book reading and materialistic design, by going back to book-binding.

◆ 1990年在杉浦老师的工作室学习
Studied in the studio of Kohei Sugiura, 1990.

◆ 在21年后的2011年拜访杉浦老师
Visiting Kohei Sugiura, 2011

杉浦康平

日本平面设计界的巨匠,著名书籍设计家,视觉信息设计的建筑师,亚洲图形研究学者,日本神户艺术工科大学教授,亚洲设计研究会会长。在85年的生命历程中,设计生涯长达半个多世纪。有着无尽的创想活力。因为博学与专注,在音乐、字体、网格、东方图形、曼陀罗、噪音学说、信息图表等方面均有研究。他的书籍设计理论领先于时代,成就斐然,在日本独树一帜,并影响亚洲几代设计人。

Kohei Sugiura

By the age of 85, he had spent more than 50 years on book design. He is a man of infinite idea of creativity. He has a high level of knowledge, shrewdness and concentration. Ahead of others, he has been leading in diverse fields, including music, typography, oriental figures, Mandala research, noise theory, information graphic and book design theory. He is unrivalled in Japan and give influence to generations of Asian designers.

◆ 即将出版的杉浦康平新著
《脉动的书——杉浦康平的设计哲学与手法》
《时间地图——杉浦康平的信息图表设计》
Vibrant Books: Methods and Philosophy of Kohei Suginrais Design, Experiments in "Time Distance Map": Diagram Collections by Kohei Suginra

068   S ⇨ 230 × 170 mm
       D ⇨ 2006

**Books, Text, and
Design in Asia: Sugiura
Kohei in Conversation
with Asian Designers**

SDX Joint Publishing

亚洲的书籍、文字与设计——
杉浦康平与亚洲同人的对话

三联书店

069   S ⇨ 230 × 170 mm
      D ⇨ 2014

**Spinning: Sugiura
Kohei's Design**

SDX Joint Publishing

旋——杉浦康平的设计世界

[日] 臼田捷治 著　吕立人、吕敬人 译

三联书店

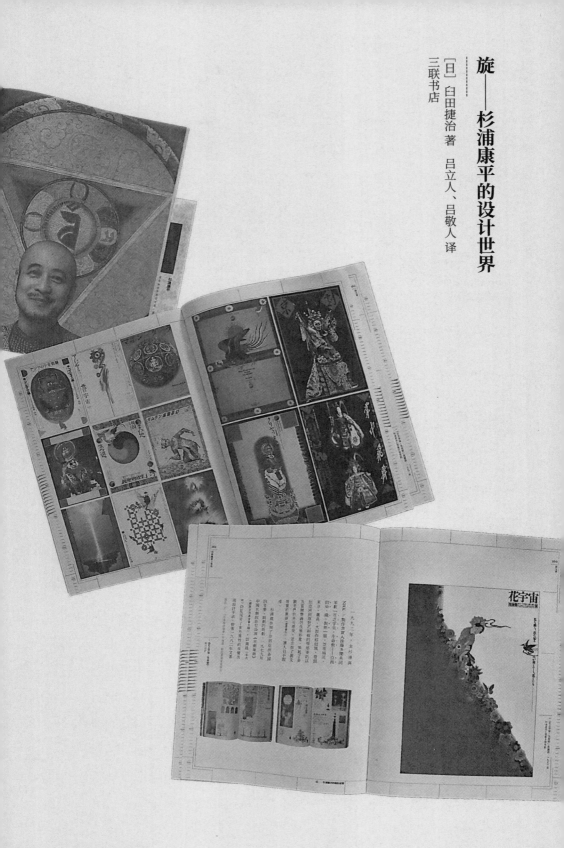

从装帧到书籍设计，这并不是对两个名词的辨识，而在于思维方式的更新，设计概念的转换，书籍设计师对自身职责的认知。从习惯的设计模式跨进新的设计思路，这是今天书籍设计概念需要过渡的转型期。时代需要以书籍设计理念替代装帧概念的设计师，从知识结构、美学思考、视点纬度、信息再现、阅读规律到最易被轻视的物化规程，突破出版业中一成不变的固定模式。装帧与书籍设计是折射时代阅读的一面镜子。

I am not going to distinguish book-binding from book design without any meaning. The two terms serve the role of opening ways of thinking towards book design and elevating the understanding level of it. The objective is to let book designers know their accountability, by escalating levels of culture and converting design concept. The concept of book design is now at a situation where stereotypical design patterns should develop into a new way of design thinking. The current age needs designers who can get over the concept of book binding via the concept of book design. This era make it an imperatie to break down knowledge structure; aesthetic thinking; viewing position; reproduction of information; tangible regulations that seem easy to observe by rules of reading; and, the sentiment that the publishing industry will remain unchanged. Book binding and book design is like a mirror that reflects the reading habit of an era.

◆ 2007年在香港设计大赛担任评委／Judge panel of Hong Kong Design Competition 2007

◆ 2012年与徐岚、韩湛宁在汕头古村落／In a Shantou ancient village, 2012

Book Design / 1998–

书籍设计

一九九八年，因为我不愿意把生命浪费在低效内耗的工作环境中，于五十岁那年离开了出版社，成立了自己的『敬人人敬设计工作室』，比『三十而立』晚了二十年。这使我在从装帧到书籍设计的观念改变和理论探索上有了更多的实践机会。面对不同的作者，别样的客户，进行着艰难的交流沟通，解释突破书衣打扮的装帧观念局限的必要性，说明同一文本可以演绎不同表情的故事的书籍设计道理。书籍设计不仅仅是一个完成信息传达的平面阶段，设计师还要拥有文本信息阅读设计的构筑意识，学会像导演那样把握在层层叠叠纸页中由时间、空间、节奏构成的语言和语法；设计书页中应该承载着知性的力量，而非漂亮的躯壳。幸运的是客户们理解了我的概念和用心，他们会在选题确定之初与我切磋该书文本内容的阅读构架，征求书籍信息视觉传播的创想点，出版人、编辑们的这种观念改变让我感到书籍设计理念正在落地生根，并对未来中国的书籍设计充满期待。

One day in 1998, I felt like that I did not want to consume my precious life any more in an inefficient and resources-exhausting system. In the year when I turned 50, I quit the job and established my own company. Judging from a Chinese classic saying, "A man should be self-supported at the age of 30", I was quite late in practicing the lesson. While running the studio, I had more chances to exercise changing thought in book binding and explore theories on it. I made exchanges and communications with various authors and clients, while feeling the need of overcoming the limits of my perception on book design to package and book binding. I also have to explain why a book design varies, even on same texts, delivering different stories. As book design does not end in conveying information in one dimension, designers have to have ways of the design construction that can get textual information. They also have to learn how to configure language and grammar through time, space and rhythm on countless pages, as a director does. Book page should be designed to be able to give the intellectual power. It should not be a simply beautiful outfit. I can truly say that I was lucky because my clients understood these stories. They talked with me when they were about to decide on book's themes and textual structures. They also shared their ideas with me about ways of visually delivering book's information. Those attitudes found in publishers and editors made me believe that book design is taking its root in China. Thanks to them, I could have expectations that the future of Chinese books is bright.

070

S ⇨ 120 × 155 mm
D ⇨ 2002

敬人书籍设计 2 号 | 电子工业出版社

《敬人书籍设计 2 号》是一本运用了 38 种纸和各种工艺做成的阅读形态十分有趣的书。一本作品集、一本评论集，两本书套合，翻阅时需要撕开纸页，并由此传出美妙的声音，在聆听中可翻看信息结构编排多样的图文。

Jingren Book Design No.2

Publishing House of Electronics Industry

This book is interesting to read, because of the 38 types of paper and diverse crafts in it. One volume contains works, and the other carries critiques. Readers have to tear off a page when they move from one page to another. In doing so, they hear the sound of paper torn and encounter diverse images and textual information.

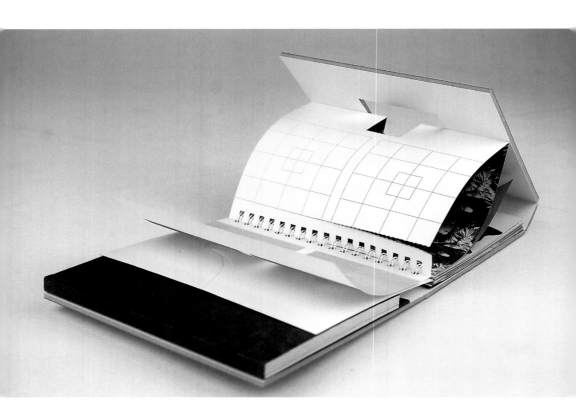

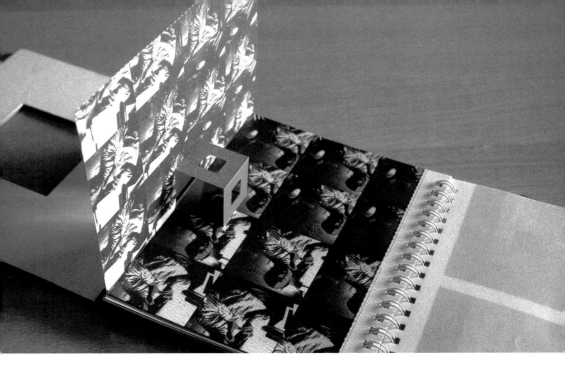

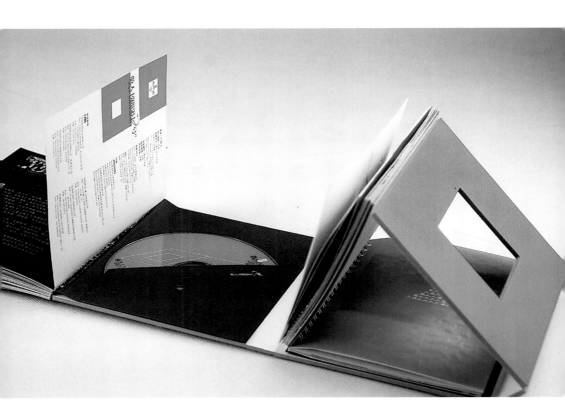

S ⇨ 140 × 210 mm
D ⇨ 2002

梅兰芳全传

中国青年出版社

原本是一本无一张图像的 50 万字纯文字的书，设计师在设计过程中寻找近百幅图片编织进字里行间，使主题内容表达更加丰满。书的切面（书口）设计为读者在左翻右翻的阅读过程中呈现出梅兰芳"戏曲"和"生活"两个生动的形象，很好地演绎出梅兰芳一生中的两个精彩舞台，为读者留下完整的印象。书籍，它不是静止的装饰之物。读者在翻阅的过程中，与书沟通，并产生了互动，书成为一个能够驾驭时空的能动的生命体。设计者不仅仅可以成为一个演员，还能是一个编剧，甚至是一个导演。

Biography of Mei Lanfang

China Youth Press

The book, initially having about 500,000 characters, was later inserted with over 100 pictures on related pages. The modification was done to enrich the content of each theme. The fore edge is designed to provide a 3D feeling, and readers may explore Mei Lanfang both in Peking opera and everyday life when they turn pages leftward and rightward, respectively. This formation of the book realistically displays the two outstanding stages of his life, so that readers could have a deep impression of him. Books are not an immovable decoration. They should communicate with readers; give-and-take feedbacks to-and-from them; and, be perceived as an active living creature that controls time and space. Editorial design is what disseminates comic information in a delicate way. A designer could be an actor/actress, producer, or director.

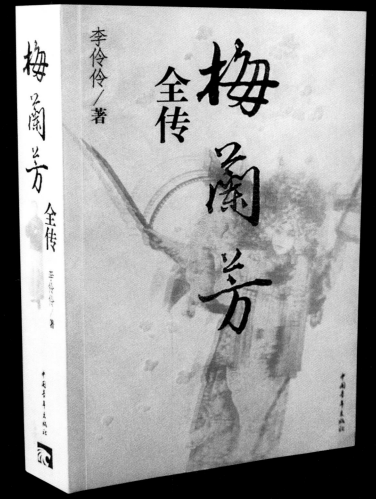

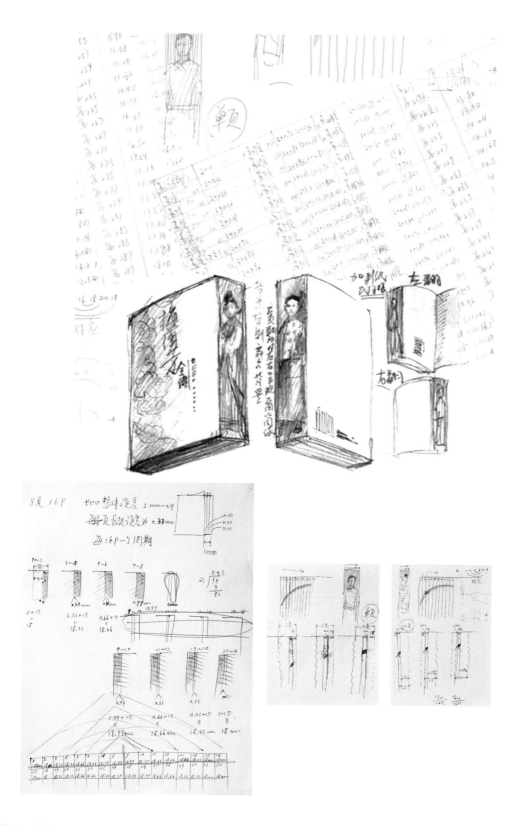

S ⇨ 155 × 260 mm
D ⇨ 2004

范曾谈艺录
中国青年出版社

《范曾谈艺录》在书籍设计上着重文本版面构成的实验性探索：文字板块在全书分布节奏的把握，字体字号行间距的阅读分列式动态变化，文字群在纸面空白中的经营游走。形态及纸质均透露着书卷气息，并特意参照中国古代文人眉批的阅读习惯，以边缘文本的极限处理为文字群的周边留出尽可能大的空间，供读者写随感之用。函盒用富有质感的手工纸打造，凸显回归自然、淡泊优雅的文人气息。

Talk on Arts-Fan Zeng

China Youth Press

Book design is focused on experimental research, by coming up with relevant text lay-outs. Texts are rhythmically placed throughout the book. The font, size and space between lines are adjusted for a feeling of moving. Characters are freely arranged on empty spaces on paper. The usage of paper forms and texture is the same with on scroll paper. The book considered the practices of ancient Chinese literati who put annotations on the top of a page. Texts are laid on one area on a page, so that the page could secure as big space as possible for reader's convenience. The case is made of hand-made paper that gives a feeling of thick and rich, as it wanted to be felt like going back to Mother Nature and becoming a plain and elegant literary person.

苦苦》。画于苦困中挣扎之坚毅少年，眉宇间有百折不挠之神采，显呈恒温儿图》。画者苦厅人之忍忱无奈，紧握爱子之手，慈母命运，纠葛缠结，生不知如何？小子泥至泥图般下，死则已吞声；生则常侧倒，至为《一篮青色卖烟人凋》。写一家贫苦女，祝牧培秀西荟春旁者，子人遗物之心酸，亦若干凋之鲜花，连慧动人。《卖小吃的老人》，或为寒寒之苦，或为睡寒之处，

旗而不鸣，就眺粉之沉于苦难
申记为地理，愚感思考
凉，亮和先生艺术之伟大，思
索就，新开低体，发于其长
笔泪成忍歌。和慈国骨中，生
际下，腐大悚颜，百年之中不乏
生一往情深，考虑花与生命活
之平凡。

先生旦作《流民图》系
华北沦陷区的北平展出一天不
禁，复及挂日落落晚年三白，劫
后于日军日仓落中失破，运至
哥捐再东北，人业后会，先生黄以
辗辙辗轮北渠。先生皇后，欣
然遭理，此画目前已藏于中国
美术馆。

蒋兆和《卖小吃的老人》

《苦苦》，我仅为统一脑之所成，其价值未有及此传作鸣篇者，深蓝悄庵
《九竹云天之《富春山居》一半得手人，游客则《流民图》列一半
完，日仿《先生之所致书画之手，欣使《流民图》完整，其则可制则其技不建。直书
《无竹气化手编事品，替给其肃事，纠不失先生《流民图》时忘气
以尽，欲定姑非同西语耳！不知其读之次流，何以其读之所以次。正
於两千余里之大作也。锐思敏悟，洞察宇宙之虑，人间两相于吾耶经
长之此上却下最是精彩，传玉代驾主命黄苦改其造于所逐钟缘绘美
於王客道："吴道子所路钟第一身之力，气色、眼神俱苦第二百。不
见人物第一颦之变争动全身。"亮和先生于此可贵脐伴
矣。

先生作画笔节月使，沈沦萧闲，目光滑溢，其笔举过之，布所敬美之尤
于缀炉，度目咯啦中指斯动人，支思因翠，诗右颜有的合态式
若此。《先生图》记，"此览篇第三叠第一部也。"此大有寿养宗澄之意
之溢迥。袁拒行之等寄实主义，亮和先生则属有高手笔色，固不其
之汪洋思然以长于斜地之把握，可漏滴水以如海海性，人物满彷光此等地秒
《流民图》，先生若于学，进于于建，其间时借远蔷薇六法，顾氏之论语，
次宗晋《芾疗学画编》皆苦推敖。先生筑聚物色精构，重
丰的形，生因其董砂悍肖，纤笔百耳，而所求万海的无之意，先生

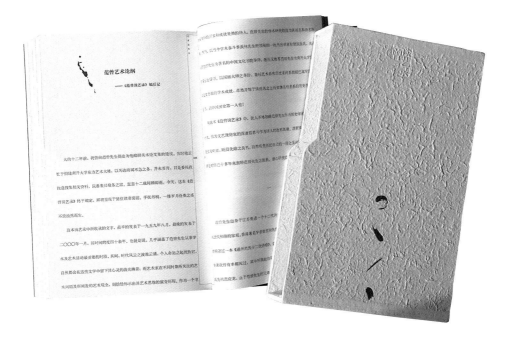

S ⇨ 140 × 180 mm
D ⇨ 2003

怀珠雅集

河北教育出版社

《怀珠雅集》是五位画家的藏书票画集。在编辑设计中将名家关于读书、藏书的只言片语编入书中，成为整本画册的组成部分，增加了阅读信息。形态既传统又求新，注入了过去文人雅士更趋向于自然、淡泊的审美意识。书名贴签周边用手撕成形后再贴在质感非常柔软的宣纸封面上，让读者能够体验到东方纸张植物纤维的亲切感。

Florilegium of Books Labels

Hebei Education Press

The book is a collection of book labels of five painters. Editorial design puts their comments in relation to reading and book collection on diverse locations in the book, thus allowing it to be a section. The form of the book is traditional yet novel, and its design offers a sense of aesthetic consciousness and closeness to nature, found in the idea of the literati. The book's title label is torn off by hand, so that it displays paper fiber. Then, it was glued to the very smooth rice paper cover. This processing was intended for readers to feel friendly to paper fiber which was once widely used in Asian countries.

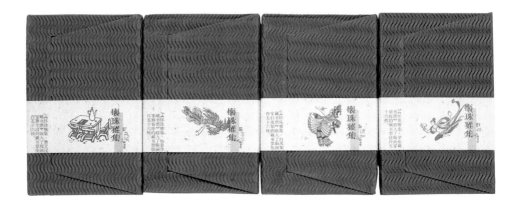

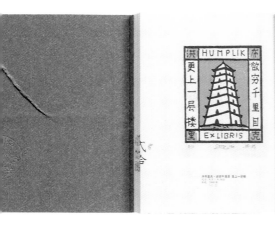

美丽的京剧
电子工业出版社

S ⇨ 165 × 245 mm
D ⇨ 2007

《美丽的京剧》以生、旦、净、丑的不同纸质为文本结构的视触分割，将传统京剧戏楼、戏牌、戏服等视觉元素渗透全书，为主题塑造浓郁的戏曲阅读氛围，编辑设计理念贯穿于作者、编辑、设计者的互动之中。

  设计强调翻阅的过程，能让读者感受进入剧场、启幕、高潮、低潮、幕间、结束的视觉氛围。如序言用红底白字，就好像是进传统戏院第一眼就能看到的戏牌；辑页分别表现生旦净丑做念唱打最典型的手势和眼神；文本格式采用20世纪30年代京剧的广告模式，传达出京剧的味道；后记页面由红色渐渐减淡，最后变成白底，意指离开剧场。版权页设计成节目单的布局。合上书页如同看完一场戏。书籍设计不是简单地把信息平面展陈在单一的页面上，一定要贯穿和调动时间和空间的信息陈述过程，注入时间、事件和story（故事）的概念。

The Beauty of Opera

Publishing House of Electronic Industry

This book considers the sense of vision and touch of paper texture in different factors, *sheng, dan, jing and chou,* and accordingly comes in diverse textures. Visual elements of opera, including theater, billboard and costumes, are well reflected in the book, so readers can feel the real atmosphere while reading. This editorial concept was the result of mutual communication among authors, editors and designers. The design of the book emphasizes the process of reading, and the whole process of opera is planned for readers to fully enjoy all stages of it (beginning, climax, slowdown, intermission and end), particularly through visual effects. The foreword page is in red and characters are in white, to remind readers of billboards of the opera that people commonly see at the entrance of theaters. *Sheng, dan, jing* and *chou* are the modes that actors and actress use to express most typical shapes of hands and eyesight. The tone of texts is borrowed from that of advertisement leaflets of the 1930s for a real feeling of Peking Opera. The latter parts of the book see the color of red weaken, before turning white

at the end. Changes in colors give a feeling of an opera coming to an end. In addition, the publication rights feels like a brochure of a performance. In all, the book is designed for readers to feel that they watched a Peking opera. Book design is not a simple process putting information on paper. It is a describing process of the information that flows through time and space. This is why book design should have the concept of time, event and story within itself.

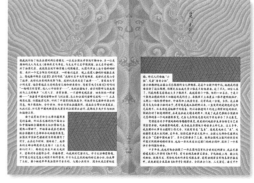

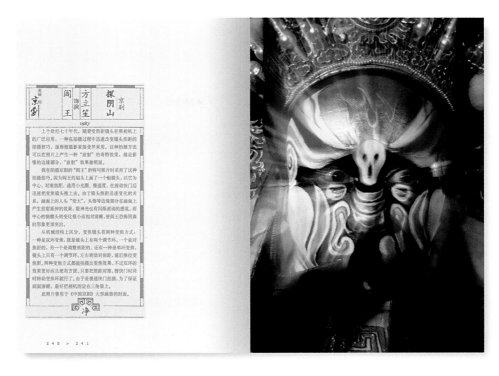

京剧 探阴山
方立笙 饰演
阎王
1987

上个世纪七十年代，随着变焦新镜头在照相机上的广泛应用，一种在拍摄过程中迅速改变镜头焦距的拍摄技巧，逐渐被摄影家接受并采用。这种拍摄方法可以在照片上产生一种"放射"的奇特效果。越是影像慢的边缘部分，"放射"效果越明显。

我在拍摄京剧的"阎王"的特写照片时采用了这种拍摄技巧。因为阎王的脸上画了一个骷髅图案，以它为中心，对准焦距，选用小光圈，慢速度，在按动快门后迅速把变焦镜头推上去。由于镜头焦距迅速变化的关系，画面上的人头"变大"。头饰等边缘部分在画面上产生放射延伸的效果，眼神光也有闪烁流动的感觉，而中心的骷髅头则变化即小相对清晰，使阎王恐怖阴森的形象更加突出。

从机械结构上区分，变焦镜头有两种变焦方式：一种是双环变焦，就是镜头上有两个调节环，一个是对焦距的，另一个是调整焦距的。还有一种是单环变焦，镜头上只有一个调节环，左右转动对焦距，前后推拉变焦距。两种变焦方式都能拍摄出变焦效果，不过双环的效果要更好而且使用方便，只要把焦距对准，按快门时同时转动变焦环就行了。由于是慢速快门拍摄，为了保证画面清晰，最好把相机固定在三角架上。

此照片曾用于《中国京剧》大型画册的封面。

净

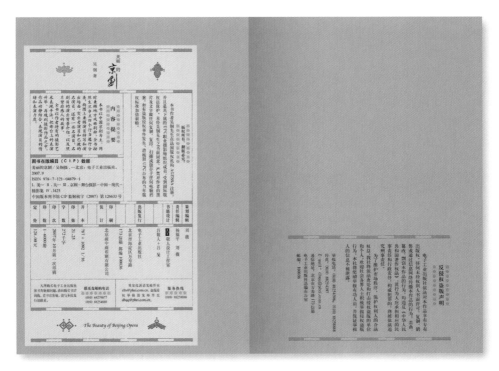

S ⇨ 185 × 260 mm
D ⇨ 2007

书戏——当代中国书籍设计家40人
南方日报出版社

《书戏——当代中国书籍设计家40人》是一本收录40位当代中国书籍设计家作品的作品集。书中设计运用有趣的戏剧化编织信息手法，把书做成一个值得玩味的纸载体。外包封通过折叠，由一本蜕变成了两册。全书用了十种纸，儿时游戏"翻花绳"的图形贯穿每一章节，将40人的作品有层次地编排出来，演绎承载不同风格作品有节奏变化的书戏舞台。这是一本贯穿书籍设计概念的自编自导自演的书。

A Play of Book-40 Contemporary Chinese Book Designers

Nafang Daily Press

This book, involving 40 leading contemporary book designers, is made in interesting ways. The book cover is double-folded, so readers could feel that they read two books. More than 10 different kinds of paper were used, and each page is linked to another in the way of a cat's cradle. Their work pieces were serially organized in such a way that each different work rhythmically appears on page. The book was the first of its kind, in that each individual designer wrote a script, directed a play and acted in it. It opened a new horizon of book design.

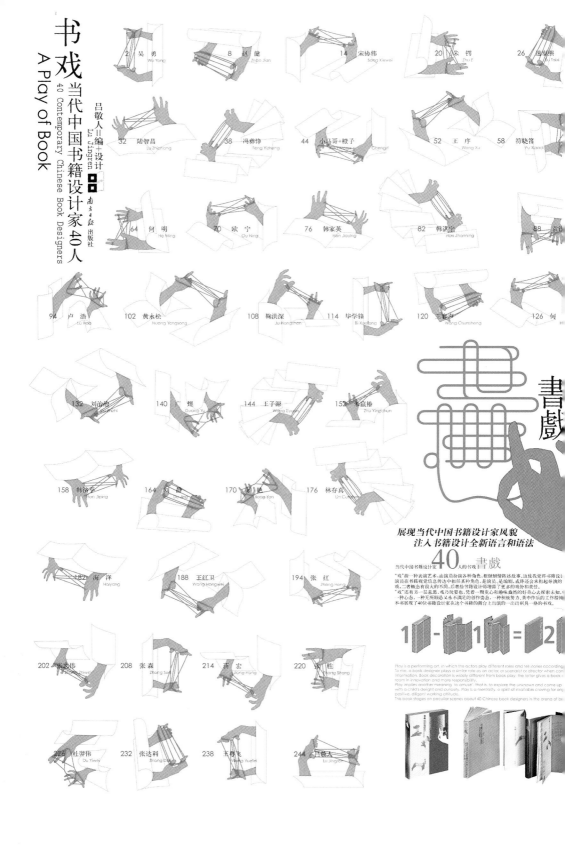

S ⇨ 170 × 260 mm
D ⇨ 2007

灵韵天成、蕴芳涵香、闲情雅质

中国轻工业出版社

《灵韵天成》《蕴芳涵香》和《闲情雅质》是一套介绍绿茶、乌龙茶和红茶的生活休闲类图书。出版社对它的定位是时下流行的实用类的快餐式的畅销书。我在与著作者的接触中为作者对中国茶文化热切投入的精神所感动，觉得书的最终形态不应该只是纯商品书物，应该让全书透出中国茶文化中的诗情画意，这也是对中国传统文化的一种尊重。这一编辑设计思路经过与作者沟通取得共识。与出版社就文化与市场、成本与书籍价值进行了反反复复的探讨，这一方案最终也得到了出版人的认可。全书完全颠覆了原先的出书思路，用优雅、淡泊的书籍设计语言和有节奏的叙述结构诠释主题。绿茶、乌龙茶两册采用传统装帧形式，内文筒子页内侧印上茶叶局部，凭借油墨在纸张里的渗透性，使得在阅读中呈现出茶香飘逸的感受；另外一册红茶从装帧形式到内文设计均为西式风格，体现英式的茶饮文化。全书没有任何矫饰和刻意的设计，但处处能让读者体会到设计的用心。此书的出版给出版社带来从未有过的书籍面貌。虽然书的价格成本比原来默认的高了些，但书的价值得到了全新的体现。

Natural Aura, Lasting Fragrance, Elegant Leisure

Chinese Light Industry Press

The books are about tea culture. They are written in simple, plain, implicit and toned-down ways, so that readers can enjoy their reading over a cup of tea. The three books have different ways of reading. The part about green tea and Oolong tea uses double-layered paper. Specifically, the lower layer is printed with the image of tea leaves for the two-level paper to offer a feeling of mist. Reader could feel that they smell the clean and pure scent of tea, oozing out of the paper.

As each of the books is about green tea, Oolong tea and black tea, the publisher originally wanted to make them like a trendy, practical and living-room book. Let's call their intention, "fast-food style best seller." While talking with the author, I was touched by the passion for the Chinese tea culture of the writer. I thought that the book should not be like any other books, but contain poetic sentiment included in the country's tea culture. It was a respect to the traditional Chinese culture. This was why I made long talks with the author to build a consensus on book design. Since then, I

talked with the publisher about tea culture and markets, costs and price of the book. Regarding the overall design, I totally changed my initial plan. Instead, I came up with a design that is done in plain and simple design languages and a rhythmical way of description, to tell the story of the theme. For green tea and Oolong tea, I employed traditional bindings and double-layered paper which had images of tea leaves between the top and the down layer. I wanted to give a sense of pluming tea scent while readers enjoying their reading, by taking advantage of the features of ink permeated on paper. For black tea, I decorated the binding and the design of main contents in Western styles, in order to talk about English tea culture. Throughout the process, I carried out my design in such way that does not give a feeling of being artificial or intentional. But I did it so that readers could feel the efforts put in the design. The publication of the book was an unprecedented trial. Though the production cost of the book was higher than expected, I created a new value of the book.

077

S ⇨ 210 × 285 mm
D ⇨ 2012

红旗飘飘——20世纪主题绘画创作研究

人民美术出版社

Let Red Flat Flow-Research on the 20 Century Theme Painting

People's Fine Arts Press

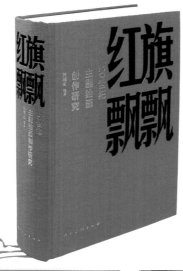

079

S ⇨ 185 × 260 mm
D ⇨ 2009

西方现代派美术

中国建筑工业出版社

Western Modernist Art

China Construction Industry Publishing House

078

S ⇨ 210 × 285 mm
D ⇨ 2005

首都博物馆常设展图录

北京出版社

Selected Art Works of Capital Museum, China

Beijing Publishing

080

S ⇨ 185 × 290 mm
D ⇨ 2008

嘉业堂志

国家图书馆出版社

History of Jia Yetang

National Library Press

081

S ⇨ 130 × 185 mm
D ⇨ 2009

毛泽东箴言

人民出版社

Quotations from Mao Zedong

People's Press

082

S ⇨ 170 × 240 mm
D ⇨ 2010

萧红全集

黑龙江大学出版社

The Complete Works of Xiao Hong

Hei Long Jiang University Press

084

S ⇨ 213 × 285 mm
D ⇨ 2008

奇迹天工

文物出版社

Exhibition of Ancient Chinese Inventions and Artifacts

Cultural Relics Press

083

S ⇨ 185 × 260 mm
D ⇨ 2008

浪漫与美丽

中国戏剧出版社

Beauty in Romance

China Drama Press

085

S ⇨ 140 × 190 mm
D ⇨ 2003

中国民间文化遗产抢救工程普查手册

高等教育出版社

The Rescue Project of Chinese Mandarin Folk Heritage Handbook

Higher Education Press

087

S ⇨ 260 × 350 mm
D ⇨ 2006

闫平画集

人民美术出版社

Album of Yan Ping

People's Fine Arts Press

086

S ⇨ 290 × 420 mm
D ⇨ 2011

老开明国语读本全系列

中国青年出版社

The Full Series of Old Kaiming Chinese Text Books

China Youth Press

088

S ⇨ 185 × 260 mm
D ⇨ 2011

毛利 A-Z
译林出版社

Mori A-Z
Yilin Press

090

S ⇨ 210 × 280 mm
D ⇨ 2015

小鲤鱼跳龙门
大象出版社

Carps Jumping over the Dragon Gate
Daxiang Press

089

S ⇨ 210 × 285 mm
D ⇨ 2005

中国广告图史
南方日报出版社

Pictorial History Of Chinese Advertisement
Nanfang Daily Press

091

S ⇨ 285 × 350 mm
D ⇨ 2011

王式廓 1911-1973
中国青年出版社

Wang Shikuo 1911-1973
China Youth Press

092

S ⇨ 230 × 295 mm
D ⇨ 2005

中国传统体育
首都师范大学出版社

Traditional Events of Sports in China
Capital Normal University Press

装帧与书籍设计是折射时代阅读的一面镜子
Book Binding and Book Design is a Mirror That Reflects the Reading habit of an Era

装帧与书籍设计的区别

装帧与书籍设计概念的区别是什么？由于特殊原因，长期以来在中国装帧只是封面设计的代名词，或仍然停留在书籍装潢、装饰的层面，即为书籍做打扮的接口。这并不排除部分装帧者对书进行整体运筹的特例，但多数的装帧则以二次元的思维和绘画式的表现方式完成书的封面和版式。其原因有三：一是设计者受装帧观念制约，把自己的工作范围限定在给书做外包装，很少去注意内文的视觉传达规律研究和书籍整体阅读架构的设计思考；二是出版也是一种产业，出版人为了控制成本，认为从封面到内文的整体设计会增加成本，影响经济效益，故并不积极主张设计师对书进行整体设计的投入；三是大部分文字编辑的专业观念还停留在把握文字质量的工作层面，却缺少对书籍信息的传达和艺术表现力的索求和愿望。这就造成目前从出版人到编辑，从设计师到出版发行人员仍然模糊地习惯于"美化书衣，营销市场"的这一"装帧"概念。

其实中国过去有许多优秀的前辈设计家并不满足只为书籍做打扮的工作接口，他们排除各种困难，创作出大批经典的传世之作。但无奈的是那时的社会环境、经济条件、出版体制、观念意识等诸多因素，并不能使设计师充分发挥他们的才智和创造力，更由于装帧原意中装潢加工的解读，而无法注入全方位的整体设计理念，仅仅停留在增加吸引力和艺术化表现层面，致使他们的创意价值的认同至今得不到完善的兑现。中国改革开放以来，新的信息载体传播态势已

BOOK DESIGN 书籍设计

③ + ①

Book Binding 装帧

Typography Design 编排设计

Editorial Design 编辑设计

要求改变这一局面，首先要改变观念，认识到装帧概念的时代局限性，作为书籍设计者，与文本著作者一样，是书卷文化和阅读价值的共同创造者，他们一定能以新的理念，付出心力和智能，展现出书籍艺术的阅读魅力。

我以为书籍设计（Book Design）包含三个层面：装帧（Book Binding）、编排设计（Typography Design）、编辑设计（Editorial Design）。书籍设计真正的含义应该是三位一体的整体设计概念。装帧只是完成书籍设计整个程序中的一个部分或最后一个阶段。"装帧"与"书籍设计"

无论是概念性质、设计内涵、涉及范畴、运行程序、信息传达、形态架构、工作体量……两者均有着很大的不同。

书籍设计应该是一种立体的思维，是注入时间概念的塑造三维空间的书籍"建筑"。其不仅要创造一本书籍的形态，还要通过设计让读者在参与阅读的过程中与书产生互动，从中得到整体的感受和启迪。那种以绘画式的封面装饰和固化不变的正文版式为基点的装帧，只是一个外包装。

书籍设计应是在信息编辑思路贯穿下对封面、环衬、扉页、序言、目次、正文

Infography Design 信息设计

书籍设计师的角色与责任

一本书的设计虽受制于内容主题，但绝非是狭隘的文字解说或简单的外包装。设计者应从书中挖掘深层含义，寻觅主体旋律，铺垫节奏起伏，在空间艺术中体现时间感受；运用理性化有序的规则驾驭、捕捉住表达全书内涵的各类要素——到位的书籍形态、严谨的文字排列、准确的图像选择、有时间体现的余白、有规矩的构成格式、有动感的视觉旋律、准确的色彩配置、个人化的纸材运用、毫厘不差的印刷工艺；寻找与内文相关的文化元素，升华内涵的视觉感受；提供在使用书籍过程中启示读者联想的最为重要的"时间"要素和对书籍设计语言的多元运用；最后达到书籍美学与信息阅读功能完美融合的书籍语言表达。这近乎是演绎一出有声有色的充满生命力的戏剧，是在为书构筑感动读者的书戏舞台。

书籍设计应该具有与文本内容相对应的价值，书应成为读者与之共鸣的精神栖息地，这就是做书的目的。一本设计理想的书应体现和谐对比之美。和谐，为读者创造精神需求的空间；对比，则是营造视觉、触觉、听觉、嗅觉、味觉五感之阅读愉悦的舞台；好书，令人爱不释手，读来有趣，受之有益。好书是内容与形式、艺术与功能相融合的读物，最终达到体味书中文化意韵的最高境界，并为你插上想象力的翅膀。

书籍设计者与装帧者的不同之处，在于设计师要了解自己承担的新角色，更增添了一份可视化信息传达的责任，多了一道综合素质修炼的门槛。书籍设计师除了要提高自身的文化修养外，还要努力涉足其他艺术门类的学习，如目能所见的空间表现的造型艺术，耳能所闻的时间表现的

体例、传达风格、节奏层次，以及文字图像、空白、饰纹、线条、标记、页码等内在组织体，从"皮肤"到"血肉"的四次元的有条理的视觉再现。书籍设计者要领会对文本进行从整体到细部、从无序到有序、从空间到时间、从概念到物化、从逻辑思考到幻觉遐想、从书籍形态到传达语境的表现能力。这是一个富有诗意的感性创造和具有哲理的秩序控制过程。

音调艺术，同时感受在空间与时间中表现的拟态艺术。书籍设计还是包含着这三个艺术门类特征的创作活动。

从装帧到书籍设计

从装帧到书籍设计，这并不是对两个名词的识辨，而在于思维方式的更新，设计概念的转换，书籍设计师对自身职责的认知。从习惯的设计模式跨进新的设计思路，这是今天书籍设计概念需要过渡的转型期。时代需要以书籍设计理念替代装帧概念的设计师，从知识结构、美学思考、视点纬度、信息再现、阅读规律到最易被轻视的物化规程，突破出版业中一成不变的固定模式。装帧与书籍设计是折射时代阅读的一面镜子。

不空谈形而上之大美，不小觑形而下之"小技"，东方与西方、过去与未来、传统与现代、艺术与技术均不可独舍一端，要明白融合的要义，这样才能产生出更具内涵的艺术张力，从而达到对东方传统书卷文化的继承拓展和对书籍艺术美学当代书韵的崇高追求。

Difference between *book binding* and book design

What is the difference between book binding and book design? Historically, book binding was another name of a book cover, or decoration of a book. Of course, limited numbers of those who made book binding took part in the overall planning of books, but an overwhelming majority of them conducted jobs of making a book cover or layout, employing ways of two-dimensional thinking and painting expression.

There are three reasons for that. First, designers limited themselves to the scope of their mission. They simply thought about making a cover, not considering an overall reading structure for a visual text delivery. Second, publishers wanted to reduce the expenses of designing a book cover and main texts for improving bottom lines. They simply focused on economic burden, without calling for higher overall design levels. Third, the concept of editing texts was about improving texts themselves, not handling ways of information delivery and artistic expression of books. These are the main reasons why the concept of book binding was restricted "to decorating books beautifully to sell in a market."

Indeed, Chinese designers in former times created works, significantly enough to be passed down to next generations. They did that even in the face of diverse difficulties, while carrying out book decoration jobs. But there were many constraints, including social environment, economic conditions, publishing systems, ideology and consciousness, so designers could not fully tap their talent and creativity. On top of that, book binding was simply understood as book decoration or book processing. They could not pay attention to an overall design scheme. As they ended up in simply drawing more interest or showing more artistic ways of expression, the value of designers' creativity has not been properly recognized. With the advent of new delivery media of information after the reform and opening-up of China, book design has seen new types of requests and trials coming, which were relevant to situational changes of the time. To respond, we have to change conventional ideas. We must know the problems of the concept of book binding, and book designers should be a value-creator for reading materials as a writer. Such a book designer would be able to show the appealing factors of book art, making use of their efforts and wisdom.

I think book design has three areas: book binding, typography and editorial design. It can be actually realized when the three become one to build a holistic design. Book binding is just a part, or the last section of all processes of book design. There are various factors that tell book binding from book design. Examples are conceptual features, meaning of design, related scopes, operational orders, delivery of information, structure of forms and workloads.

Book design requests dimensional ways of thinking, as book-making is a process of building a three dimensional space where time concept stays. Designers make forms of books, but also partake in the process of reading. They also have to interact with books, obtaining appreciation and enlightenment in the interaction. The book binding that is based on painting-styled book covers and stereotyped typography of main texts are no more than a simple package.

Book design should edit book's information in accordance with ways of thinking. Book covers, empty pages between covers and main texts, title page, foreword page, table of contents, forms of main texts, delivery method, rhythm and beat, character graphic, empty space, decoration, line, marking, page numbers. All intrinsic elements should be orderly and visually expressed in four dimensional ways. Book designers have to check virtually everything related to books. For example, overall picture and each detailed item; things in disorder and in order; space and time; conceptual and tangible elements; logical and illusive factors; forms of books ;and, expression ability of linguistic environments. This effort leads to the creation of poetic emotion and the process that adjust philosophical orders.

Roles and responsibility of book designers

When designing one single book, there could be constraints on contents and themes. But still book design should be neither a simple explanation made for texts, nor package. Designers should know implications and meanings of a book and, accordingly, create a space and express a feeling of time based on its theme and rhythm. They have to thoroughly capture varied elements, including: rational and orderly regulations; forms of books and an elaborate typography and selection of correct images; empty spaces where the expression of time is vivid; regular formality; rhythmical and visual melody; clear color lay-outs; use of unique paper; and printing crafts without error. Understanding cultural elements related to texts and expressing an implicit, visual feeling in a plentiful way are also the responsibility of book designers. They have to make readers think of time elements and make use of book design languages in various ways. Last but not least, the aesthetics of books and the functions of reading information should be expressed, completely. The book design that secures aforementioned factors will be very interesting, in that it sets up the stage of books that moves readers.

Book design should include values that correspond with the main contents. A book should be a resting place where readers feel comfortable, and this is the reason why we make it. An ideal book must be able to express the beauty of a harmonious contrast. The "harmonious" means that providing a mentally needed place for readers, while the "contrast" indicating that building an interesting stage of reading that appeals to five senses: seeing, touching, hearing, smelling and tasting. A good book hardly gets off from readers' hands. The more a reader find it interesting to read a book, the more the book gets valuable. A decent book fuses contents and formats, and artistic and functional aspects, providing readers with sufficient values. It also helps readers to strengthen their imagination, by allowing its implicit culture to reach an ultimate level.

If and when there is a difference between a book designer and a book binding maker, the designer has the responsibility to understand given roles and deliver them in a visual manner. In addition, he or she has to train him- or herself to increase comprehensive knowledge. He or she also must study diverse arts, for example, of spatial expression; time expression; and mimesis arts that can be felt within space and time. Book design reflects characteristics of the three areas, in

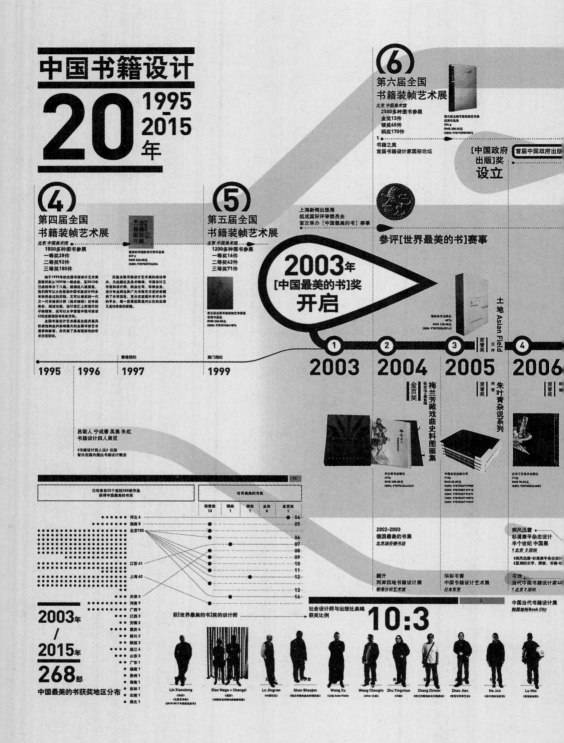

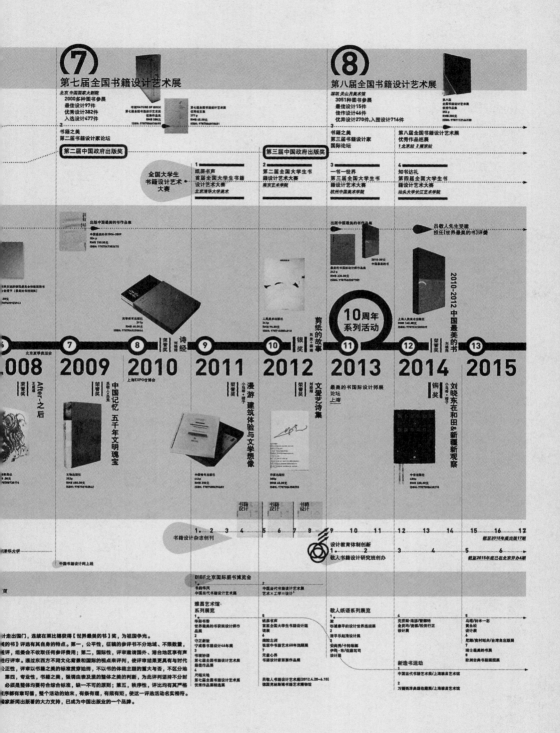

accordance with different types, in order to create creative works.

From book-binding to book design

I am not going to distinguish book-binding from book design without any meaning. The two terms serve the role of opening ways of thinking towards book design and elevating the understanding level. The objective is to let book designers know their accountability, by escalating levels of culture and converting design concept. The concept of book design is now in a situation where stereotypical design patterns should develop in a new way of design thinking. The current age needs designers who can get over the concept of book binding via the concept of book design. This era is bound to break down knowledge structure; aesthetic thinking; viewing position; reproduction of information; tangible regulations that seem easy to observe by rules of reading; and the sentiment that the publishing industry will remain unchanged. Book binding and book design is like a mirror that reflects the reading of an era.

I am not talking about the metaphysical beauty. While paying attention even to "minor techniques" of physical things, we have to walk the fine line between the East and the West; the past and the future; tradition and modern; and, arts and technology. Only when we comprehend the need of fusion and its meaning, the power of arts within book design works can exert its strength. This materialized influence can expand the traditional book culture and pass it down to our children and their children, pursuing the noble value of contemporary books that is directed to the aesthetics of book art.

093

S ⇨ 165 × 245 mm
D ⇨ 2005

烟斗随笔

国际文化出版公司

Essay of Pipe

International Culture Press

094

S ⇨ 190 × 245 mm
D ⇨ 2011

文化与价值 —— 世界美术简史

中国青年出版社

*Culture and Values-
A Survey of the Humanities*

China Youth Press

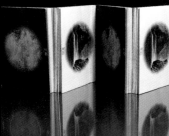

095

S ⇨ 170 × 230 mm
D ⇨ 2002

对影丛书

河北教育出版社

DuiYing Series·the Style of Writing and Painting

Hebei Education Press

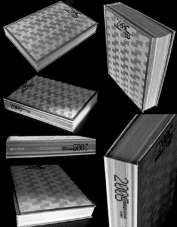

096

S ⇨ 210 × 285 mm
D ⇨ 2005

2005中国装帧艺术年鉴

中国统计出版社

Annals of China Book Design

China Statistics Press

097

S ⇨ 290 × 420 mm
D ⇨ 2011

枕边书香

北方红星文化艺术公司

The Pillow Book

Red Star Culture Company

098

S ⇨ 290 × 420 mm
D ⇨ 2011

华夏意匠——中国古典建筑设计原理分析

天津大学出版社

Cathay's Idea-Design -Theory of Chinese Classical Architecture

Tianjin University Press

099

S ⇨ 275 × 280 mm
D ⇨ 2010

中国2010年上海世博会官方图册

中国出版集团公司

EXPO 2010 Shanghai China Official Album

China Publishing Group

100

S ⇨ 290 × 420 mm
D ⇨ 2011

走进宁波

中央编译出版社

Panorama of Ningbo

Central Compilation & Translation Press

101

S ⇨ 290 × 290 mm
D ⇨ 2003

邵华将军舞蹈摄影艺术

中国摄影出版社

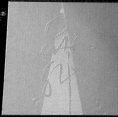

Collection of Shao Hua's Photography of Dancing

China Photography Press

102

S ⇨ 210 × 285 mm
D ⇨ 2014

民国时期电影杂志汇编

北京图书馆出版社

Collection of Film Magazines in Minguo Period

National Library Press

103

S ⇨ 150 × 230 mm
D ⇨ 2003

外交十记

世界知识出版社

Talks on Diplomacy

World Knowledge Publishing House

104

S ⇨ 180 × 270 mm
D ⇨ 2003

墨香红楼

北京图书馆出版社

Ink Fragrance, A Dream in red Mansions

National Library Press

105

S ⇨ 200 × 290 mm
D ⇨ 2009

杂碎集

上海人民出版社

He Youzhi's Collection of Essays

Shanghai People's Press

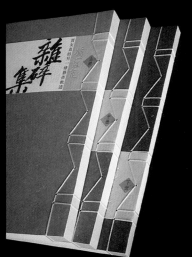

106 S ⇨ 370 × 380 mm
D ⇨ 2005

刘宇廉的艺术世界 — 黑龙江美术出版社

Liu Yulian's Art World

Heilongjiang Art Press

S ⇨ 215 × 290 mm
D ⇨ 2008

李冰冰 十年·映画

青岛出版社

《李冰冰 十年·映画》原本是出版社为李冰冰出的纯写真集，没有文本。设计者在编辑设计的过程中增添了记述她成长经历的文字，为书增添了更丰富的内涵。书做成两本，一本是她的写真，一本是她的文字，采用对折的形态。上本为黑白的小照片加文字，下本是照片，当两个页码对应起来时，照片由16开变成了8开，加大了视觉冲击力。

Impression: Li Bingbing's Ten Years Impression

Qingdao Press

The publisher originally planned to publish Li Bingbing's photo book. As there was no text regarding the background of her childhood, the book was divided into two volumes; one for photos, and the other, which is foldable, for writings. The first is only with images, while the second was small-size black and white images with comments. When you compare a page on the first with another on the second, both being on the same page number, you will see the picture of the former is twice bigger than that of the latter. The greater image is visually influential, while the texts accompanying the smaller picture giving inner meanings in a detailed way.

S ⇨ 210 × 290 mm
D ⇨ 2008

诗韵华魂
陕西师范大学出版社

《诗韵华魂》共分六卷，汇编了自先秦至近代以来的优秀诗词歌赋作品。封面烫印书法字标题，分别辅之以淡雅的传统灰色调纹样。油墨印刷于柔软而富有粗犷肌理的特种纸张上，既形成此套书的视觉特征，又体现了诗歌凝练唯美、或言志或抒情的特质。每一卷的纹样分别提取自相应时代的典型织物和器皿的装饰纹理，同时在设计上进行归纳以形成六卷密度均衡的纹理。经过不同的对比配色与题名的点缀更增加了系列书整体庄重的感觉。内文设计结合少量中国传统元素，隔页则进而延续了封面的色彩纹理与肌理质感，使读者可以充分感受这套书带来的阅读气氛。

Shi Yun Hua Hun

Shaanxi Normal University Press

This is a series book of six volumes, carrying the collection of outstanding poets from the Qin Dynasty to modern China. The title of the book in calligraphic characters is embossed as other parts of the patterned paper are pressed downward. The dimensional effect is more supported by gray-tone traditional elegant patterns on the paper. Ink printing was done on soft yet rough texture paper. Keeping its identity in mind, the book visually displays features of poetry, including simple and aesthetic beauty, implication and emotion. Patterns of a volume are representative fabrics and living wares of the era that appear on the volume. Regarding selection, patterns that commonly appear on each and every volume of the collection were chosen for readers to possibly know that they are a series book. Different colors were picked up so that readers tell one from others, not to mention of a feeling of variety that is found in any collection book.

The design of main texts combines diverse traditional elements. Inside pages used the color patterns and textures of the cover page, in order for readers to know that they are reading series books.

109 　　S ⇨ 210 × 285 mm
　　　　D ⇨ 2009

革命的时代——
延安以来的主题创作研究
人民美术出版社

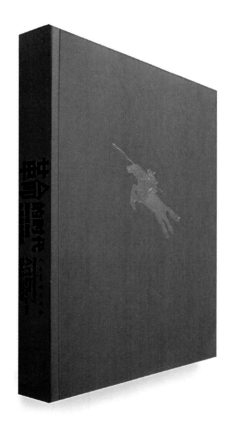

Revolutionary Art
Since the Yan'an Era
1942-2009

People's Fine Arts Press

S ⇨ 185 × 250 mm
D ⇨ 2011

剪纸的故事

人民美术出版社

《剪纸的故事》重新演绎了一位剪纸演出者的作品。在编辑设计的过程中将剪纸的味道、民间的艺术、现代的概念融入书中，运用了多种色彩丰富的、柔软的纸和彩色的线，分解重构文本作品，使作者的剪纸艺术游走于纸本之中。模拟由外往里剪纸的方式，呈现许多断页，上下拼合分离，产生灵动。书中有两张最薄、最柔软的纸，象征民间窗花用纸，展开后，是一幅很大的艺术作品。有意将机台上裁切下来的纸屑装入每一本书的塑封里，读者取出书，纸屑纷纷扬扬洒落下来，似乎是刚完成的剪纸作品，还带着演出者的温度。该书获2012年"世界最美的书"银奖。

The Story of Paper-cut

People's Fine Arts Press

Many considerations were made to put work pieces of paper-cut artists on stage. Pieces of soft paper in diverse colors and different lines were employed for the right understanding of the charm of paper-cut art and a good description of folk art and modern concept. By restructuring the content of texts, the cut is made outwards. This method gives a feeling that the characteristics of paper-cut are alive within the book. Pages were cut to pieces. Other pages in the combined form of the upper and the lower part were designed to divide, so that they could feel dynamic. The book has two sheets of very thin and soft paper between pages, which was to deliver the message of window decoration paper for a household. If and when spread wide, it displays a big art work. Pieces of cut-paper are put in the package of the book, so that they scatter when the buyer pulls them out. This is intended to show, convey and share an idea that a paper-cut artist just finished his or her work. The book was awarded the Silver Prize at the Best Books from all over the World in 2012.

111　　　S ⇨ 340 × 280 mm
　　　　D ⇨ 2012

中华舆图志
中国地图出版社

每一部地图的制作都充分体现了艺术与科学的巧妙结合。舆地图所表现的疆域图更是包含有山川、城镇、风物等综合、复杂的数据信息。古代中国舆图侧重形象描绘，本身具有很高的艺术价值，明清至近代，西方的测量技术与地图表现手法被更多地引进到中国舆图的设计制作中。《中华舆图志》一书设计上采用中式包背装，扁长形开本。筒子页增加了柔软的翻阅手感，传统的包背装载体与这部书所表达的内容充分契合。版式疏朗大方，书口用六种色彩划分章节，扁长的版心使长卷地图得到充分展示。内页中长短夹页设计十分独特，有的还原了原本装订于古籍中的书页，产生了书中书的效果；有的则通过3折甚至4折页来连贯展示超长卷舆图的全貌。在全书尾声的"舆图一览表"中，中国历史大事记年表和舆图绘制信息

A Collection of Chinese Antique Maps

China Map Press

Making any map is a mysterious combination of art and science. Images shown on a general map, including mountains, streams, cities, and scenery and customs, comprehensively mix to convey complicating data and information. Ancient Chinese maps put greater focus on describing images than other elements. This is how got to have high artistic values. From the Ming and the Qing Dynasty to modern times, Western surveying technique and map-making method were introduced in the design process of Chinese maps. The book design of this collection adopted the Chinese map-making style to package the book's back in a long-shaped form. The double-layered pages strengthen a feeling of soft touch. The reason for the adoption was that the traditional Chinese way was proper to deliver the content of the book. A comfortable printing, fore edge where six colors divide chapters and clauses, and the long and flat part in the middle of the printing elevated the visibility of the map on scroll. Uniquely-designed inside pages were inserted into the book. At some locations, such pages

及尺寸信息在同一时间轴与同一空间内进行比对。地图年代、体量与制作形式的区别一目了然。反过来也可根据该一览表，轻松地通过书口信息检索原图。封面采用可印刷丝制材料，回归中式传统装帧手法。柔软的书页平躺在书架上，从中华舆图古老山川、河流、城市中散发出的五彩光泽等待着地图研究者与读者们去发现。

that you can just find in ancient books were placed in, in hopes of offering a feeling that another book is within it. Certain pages were folded three, or four times, producing a long-shaped version of the map. The table at the end of the book used one single axis of time to organize China's major historic events to the manufacturing and size information of general maps at one space. The years, sizes and methods of making those maps are in good arrangement.

The list and the information on the fore-edge help readers to locate original maps very easily. The book cover used printing-available silk materials and employed a traditional Chinese book binding way. On a bookshelf, the "beam" emitted by the combination of mountains, streams, rivers and cities feel like that it wants map researchers and readers to follow it.

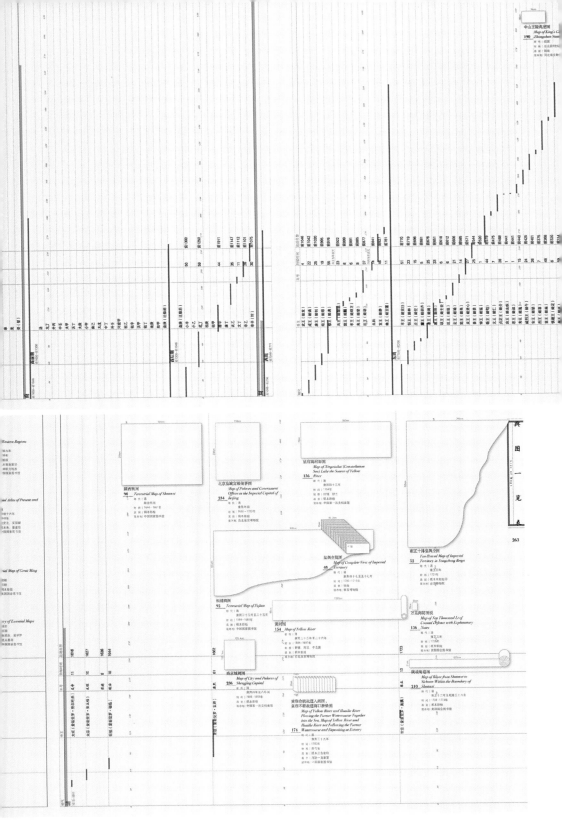

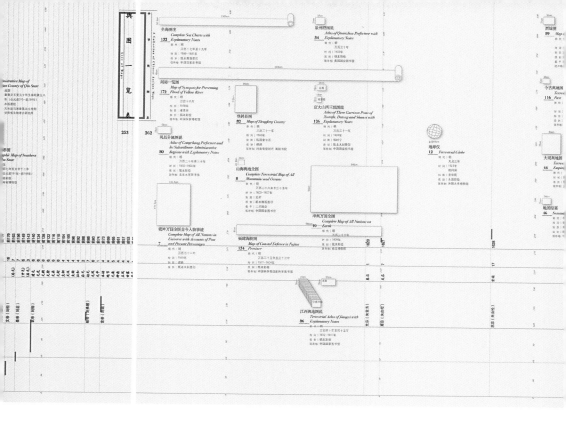
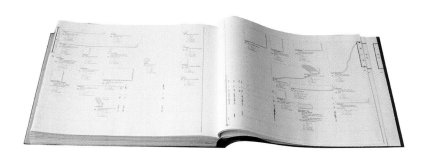

S ⇨ 170 × 170 mm
D ⇨ 2012

书筑——历史的『场』

中国建筑工业出版社

书是信息诗意栖息的建筑，东方的书籍就像森罗万象的建筑，西方的书籍犹如一座座坚固的建筑。设计师要拥有阅读设计的信息构筑意识，将空间和时间融入书籍设计概念中。随着书页的翻动，阅读体不断变换着空间关系，文本产生了时间的流动，信息通过阅读在书籍间游走，如同一栋建筑必须通过人在其中的活动才能体现并得出最佳居住审美的结构关系。

Locus -Identified by the History

China Construction Industry Publishing House

Books are a building where the poetical meaning of information dwells. Books of the East are like a construction that accept all information of the universe, while those of the West being a solid structure. Designers must have the will to establish the information of reading design. They also have to be able to dissolve space and time into the concept of book design. While reading a book, the information of it moves here and there. The same applies to a building. A building is recognized as a good residential area only when people perform their activities and live there. Otherwise, it is just a physical space.

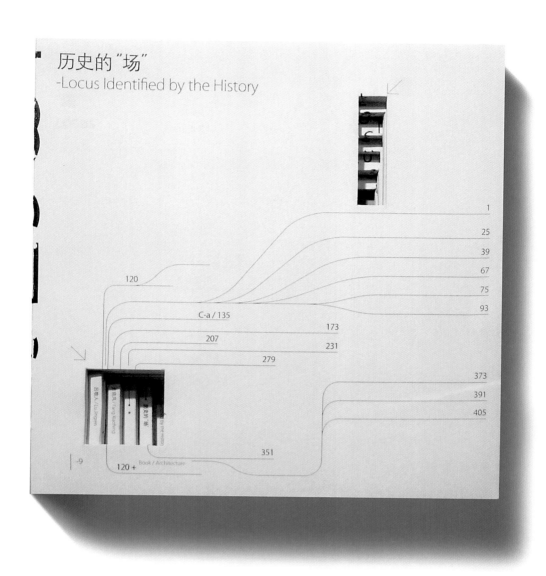

妙法自然海报 1

Ingenuity Follows Nature

114

D ⇨ 2011

妙法自然／沉郁之美 Confucianism (Chroma)

Beauty of Implicitness

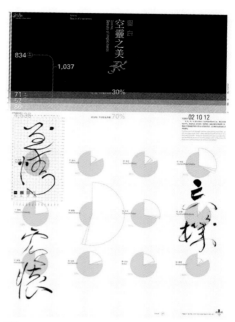

115

D ⇨ 2011

妙法自然／空灵之美 Zen (Spacing)

Beauty of Emptiness

116

D ⇨ 2011

妙法自然／飘逸之美 Taoism (Continuity)

Beauty of Ethereality

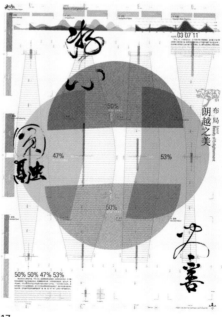

117

D ⇨ 2011

妙法自然／朗越之美 Apperception (Layout)

Beauty of Enlightenment

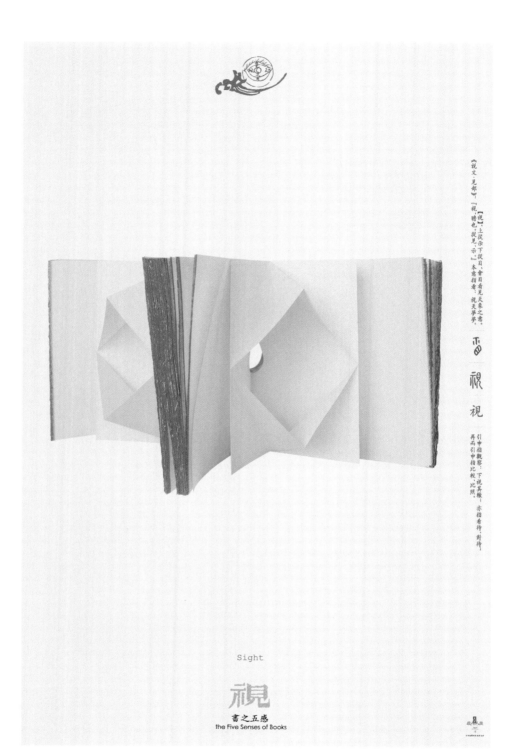

118　书之五感

D ⇨ 2005　The Five Senses of Books

Teaching Life / 2002–

教学生涯

一九九八年我成立独立工作室后，相继受邀在中央美术学院、中央工艺美术学院等各地院校，以及全国出版系统讲学。二〇〇二年正式入职清华大学美术学院，担任教学工作，并保留了自己的工作室，令其成为学生们的课外学堂。我的设计教育除了教授设计方法论，更希望打破固有的教学模式，活跃授课方法，让同学们走出校门，培养他们触类旁通的艺术审美能力和创想力。我们和世界各国以及港澳台地区的大学交流教学，请进来，走出去，教学相长，获益良多。二〇一二年我六十五岁退休，与人合作成立『敬人纸语』；二〇一三年开办『敬人书籍设计研究班』至今，与享誉国内外的书籍设计演出者共同授课。致力于当代中国书籍设计艺术的理论研究和创作实践，关注国际信息载体发展潮流。研究班通过授课、学术讲座、Workshop、手工工作坊、国外考察等多种教学方式，开拓书籍设计视野，丰富设计语法与语言，提升实际创作能力，探索新的设计教学方法，重新认知书籍未来的价值和方向。

Since I established an independent studio in 1998, I have taught students at the Central Academy of Fine Arts, the Central Academy of Crafts and Fine Arts (currently, Academy of Arts & Design, Tsinghua University) and other universities. I have also delivered academic lectures and classes to the Chinese press industry and colleges nationwide. I officially became a professor of Academy of Arts & Design, Tsinghua University, making my workshop become a place for tutoring. I gave students design methodology class, while making my teaching more active than the existing one. My objective was to help students nurture their artistic capability and creative imagination. I utilized exchange programs with other universities, including in Hong Kong, Macau and Taiwan. Through this, I harvested bumper crops, that is, the realization of mutual growth of professors and students.

In 2012, after retirement at the age of 65, I founded Jingren Paperlogue. Since the establishment of the Jingren Book Design Workshop in 2013, I have delivered classes, together with domestic and international book design artists. I am working hard for theoretical research and creative activities of the Chinese book design art, while paying attention to development trends in information technologies and media in the world. My design workshop runs a variety of educational programs, including, class, academic forum, collective training, making a hand-made book and overseas field trips. They are conceived to widen the perspective of book design and to foster the capability of actual creation through diverse design materials and languages. On top of that, I am putting in efforts to explore a new design teaching method and ideas about the future value and direction of books.

◆ 1—3
2015年在韩国坡州PaTI设计学校教学／Workshop in Paju Typography Institute, 2015

2

3

◆ 2016年和安尚秀在韩国坡州书城／With Ahn Sang-soo in Paju Bookcity, 2016

◆ 2010年在德国奥芬巴赫艺术设计学院举办讲座
Lecture in HFG University of Art and Design, Germany, 2010

◆ 2011年在日本东京MEME设计学校教学／Lecture in MeMe Design School, Japan, 2011

◆ 2007年杉浦康平在清华大学美术学院举办讲座
Sugiura Kohei's Lecture in Academy of Arts & Design, Tsinghua University, China, 2007

◆ 2011年清华大学美术学院东亚设计教育论坛
Design Education Forum of East Asia in School of Art, Tsinghua University, 2011

◆ 2011年在日本东京印刷博物馆举办讲座
Lecture in Tokyo Printing Museum, Japan, 2011

◆ 2015年在四川美术学院举办讲座
Lecture in Sichuan Fine Arts Institute, China, 2015

◆ 1—2
在清华大学美术学院教学
Teaching in Academy of
Arts & Design, Tsinghua
University

1

2

3

◆ 3—4
2012年在德国奥芬巴赫艺术设计学院做Workshop
Workshop in HFG University of Art and Design, Germany, 2012

4

◆ 2010年清华大学美术学院与韩国成钧馆大学艺术学院在首尔交流教学
Cooperative Teaching between Academy of Arts & Design, Tsinghua University and School of Art, Sungkyunkwan University in 2010

◆ 2011年清华大学美术学院师生在韩国坡州书城交流
Academy of Arts & Design, Tsinghua University's Faculty and Students Visited Paju Bookcity in 2011

1

2

◆ 1—3
敬人书籍设计研究班在韩国坡州出版城文化财团理事长金彦镐和李娜美设计工作室交流
The students of Jingren's Book Design Workshop communicated with Kim Eon-ho, is hairman of Bookcity Culture Foundation and Nami Rhee Design Office in Paju Book City

3

◆ 2009年与郑丙圭在韩国坡州书城活版工房／Visiting Movable Type Workshop with Chung Byoung-kyoo

◆ 2004年在新加坡南洋艺术学院教学
Workshop in Nanyang Academy of Fine Arts, Singapore, 2004

◆ 2007年美国帕森斯设计学院师生访问敬人设计工作室
Students and Teachers of Parsons School of Design visited Jingren Art Design, 2007

◆ 2013年奥芬巴赫艺术设计学院师生访问敬人设计工作室
Students of HFG University of Art and Design in Jingren Art Design, 2013

◆ 2014年卓斯乐教授携德国斯图加特艺术设计学院同学访问敬人设计工作室
Prof. Niklaus Troxler and his students visited Jingren Art Design, 2014

◆ 2006年香港理工大学师生访问敬人设计工作室／Students and faculty of Hong Kong Polytechnic University visited Jingren Art Design

S ⇨ 190 × 245 mm
D ⇨ 2006

中国青年出版社

书艺问道

《书艺问道》是一本解读书籍设计新观念和设计语言多重表现的教科书，解读装帧、编排设计、编辑设计、可视化信息设计等多向领域的交叉运用规律，注重工具书的秩序性、导读性等特征。

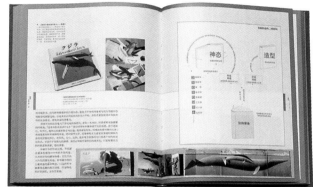

"Tao" of Book Design

China Youth Press

This is a text book that interprets diverse ways of expressing new thoughts and languages in book design. It carries principles that can be cross-applied in different processes, such as binding, typography and editorial design. There are basic principles in the book, and the design is done in consideration of the features suggested by directions.

120   S ⇨ 190 × 245 mm
D ⇨ 2006

书籍设计基础
高等教育出版社

Basics of Book Design

Higher Education Press

S ⇨ 210 × 285 mm
D ⇨ 2008

吕敬人书籍设计教程

湖北美术出版社

Curriculum of Lu Jin-gren' Book Design

Hubei Fine Arts Publishing House

书艺问道
吕敬人书籍设计说

上海人民美术出版社

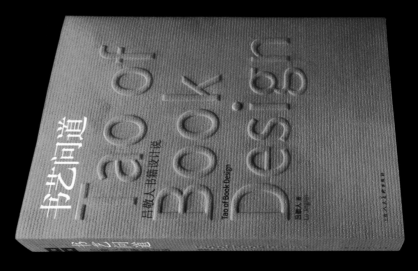

S ⇨ 192 × 245 mm
D ⇨ 2017

Tao of Book Design

Shanghai People's Fine Arts Publish House

S ⇨ 280 × 210 mm
D ⇨ 2010

敬人书籍设计研究班折纸讲义

Folded Lecture Notes for Workshop

Self-publishing book

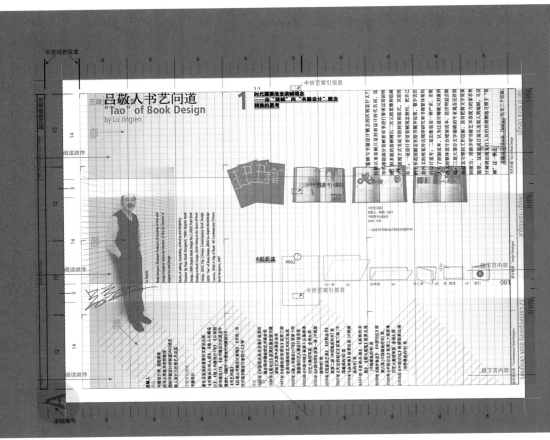

◆ 敬人书籍设计研究班折纸讲义／The Grid about Folding Text Book of Jingren Book Design Course

书籍设计教育的学术主张

长期以来书籍设计师的装帧观念只停留在二维平面构成和外在打扮的层面,使书籍设计的认知范围相对狭窄,影响设计者就文本进行有创造性的设计努力。我们提出了书籍整体设计（Book Design）的概念：完整的书籍设计要求设计师完成装帧（Book Binding）、编排设计（Typography Design）和编辑设计（Editorial Design）三个层面的工作,即"Book Design"是一个可视化信息再造的过程。

这里有两个新的概念：

一为书籍编辑设计"Editorial Design",这是书籍整体设计的核心概念,是探究信息本质并通过完成信息解构重组的文本赋予受众可视化阅读过程。书籍设计不只图封面好看,而是整体概念的完整,设计者在尊重文本准确传达的基础上,去精心演绎主题,以达到文本内涵的最佳传达效果。

二是可视化信息图表设计"Infographic Design",这是书籍整体深度设计的重要补充。设计者要掌握和分析信息本质,依循内在的秩序性与逻辑关系,构建便于受众理解的可视化信息系统,演绎出有趣、有益、有效的信息传达语言、语法和语境,以适应当今信息传达可视化的阅读需求。这是国内设计领域和设计教育尚未开发的新课题。

"书籍设计3+1"的概念缺一不可,无论是纸面信息载体还是电子书籍均可应用这一规则。Book Design是令书籍载体兼具时间与空间、兼备造型与神态、兼容动与静的信息构筑艺术。书籍的设计与其他设计门类不同,它不是一个单个的个体,也不是一个平面,它具有多重性、互动性和时间性,即多个平面组合的近距离翻阅的形式。通过眼视、手翻、心读,全方位展示书籍的魅力。书籍给我们带来享受视觉、嗅觉、触觉、听觉、味觉五感之阅读愉悦的舞台。

随着信息化时代的到来,信息传递多了电子载体这一管道,传统书籍设计虽然受到数码视频媒体的挑战,但我却认为书籍艺术设计观念终于迎来了全面更新的机会,书籍的生命价值也可赢得再次重生的千载难逢的机遇。

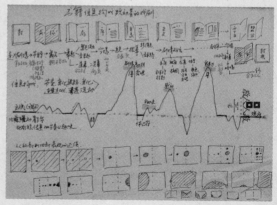

◆ 书籍信息构成跃动着的戏剧
Text information is a kind of dramatic Play

Academic Argument about Book Design Education

For many years, the meaning of book binding has stayed at two-dimensional formation and outer decoration. This narrow perception has been unchanged. Designers did not make due efforts to make creative designs for texts. According to the holistic book design concept, a perfect book design is where a designer completes three processes, book-binding, typography and editorial design. Book design is an activity that the visualizes information that already exists.

Here are two new concepts.

First, editorial design is a core concept that handles an overall design of a book. It gives a visual reading process to texts, through studying information essence and reshuffling information structure. It is not about outer decoration but about all processes related to a book. Designers have to deliver the content of texts correctly. For that, they have to make desired efforts and distribute implicit meanings of texts in the most ideal way.

Second, infographic design plays a complementary role, which is critical to design a book in a detailed way. Designers have to grasp and analyze the essence of information and build an information system for readers' easy understanding in consideration of inner order and logical relation. By providing language, grammar and linguistic environments that is interesting, productive and efficient, they have to meet the request of visual information distribution of a modern society. This is what the design field and the design education in China have so far failed to achieve.

The concept Book Design 3+1 should not miss anything. Analogue media and e-books also have to follow this rule. Book design allows the medium of a book to artistically build time and space; shape and attitude; and, dynamic and static information. Unlike any other designs, book design is neither simple nor flat. It is multiplex, mutual and diversely flat, which is influenced by time. The design is a form of short-distance reading. Through eyesight, hand-turning and reading by heart, overall charms of a book are felt. Books are a gift to us as they please our five senses, vision, smell, touch, hearing and taste.

The arrival of the information era brought digital media, a new delivery channel of information. Though digital visual media of the new age posed a challenge to the traditional book design, I find it a long-awaited opportunity through which we can improve the thought on book design. I believe this is the chance of a lifetime.

119
S ⇨ 210 × 285 mm
D ⇨ 2006

清华美院&香港理工交流教学报告书

Exchange Teaching Report of Academy of Arts & Design, Tsinghua University and Hong Kong Polytechnic University

关于网格设计的思考

20世纪80年代末,我在杉浦康平先生那里接受了网格设计的教育,令在国内只懂得装帧书衣的我茅塞顿开,引导我真正跨进设计行当的门槛,因为它不仅是设计手法,更是一种思维范式和设计的态度。他指出,设计是驾驭秩序之美的过程,网格为文本建立起一条让信息变得有效、清晰、易懂并便于记忆的控制路径,其提供了逻辑解决问题的钥匙与解开二维、三维,乃至时间传播的、并非固态的辩证方法论。网格设计是一种平面设计师必须拥有的设计意识和能力。

以西方文字构建起来的网格系统是否适用于东方汉字,我们会提出这样的质疑。西方编排网格使用的黄金比率以其严密的理性、规范科学的逻辑设定了具有普遍意义的设计范式。但中国古代各种韵文都有特定的格律,构建起语言的结构骨架,同时,方块量体的汉字更适合向量化的推算布局。杉浦老师的研究实践证明完美转换西方文字的网格理念为汉字体系所用的可行性。由此,中国的设计者完全可以应用他阐述的原理完成汉字排版网格设计过程,学会自由驾驭秩序之美的设计意识,获得构成阅读审美的自觉途径,并掌握体现东方文本韵律设计的方法论。

时代需要我们冷静面对存在的诸多不足与问题:书籍设计不是载体形式的无限叠加,应成为影响人们生活形态的一种文化行为。书籍设计者不是产出大量仿效品的出版机器,应当是文本与设计、思维与感知、思想与观念的诸多层面的文化角色的担当。设计不是为一时的炫耀,而是提供平实持久、令人感动的好作品,并服务于人们的文化生活。好书让阅读回归生活常态,阅读成为生命持续的能量库。

不可否认书籍是文化商品,既要顾及利润,更要加强有整体设计概念的产品策划和投入。美书,有引人入胜的选题文本;美书,有与众不同的阅读概念;美书,有严谨周全的逻辑秩序;美书,有物化细密的人文关照。出版人要鼓励不为一时的商业回报而精心投入的创作行为,哪怕是探索意义的作品,这才有可能满足未来电子时代的读者需求。

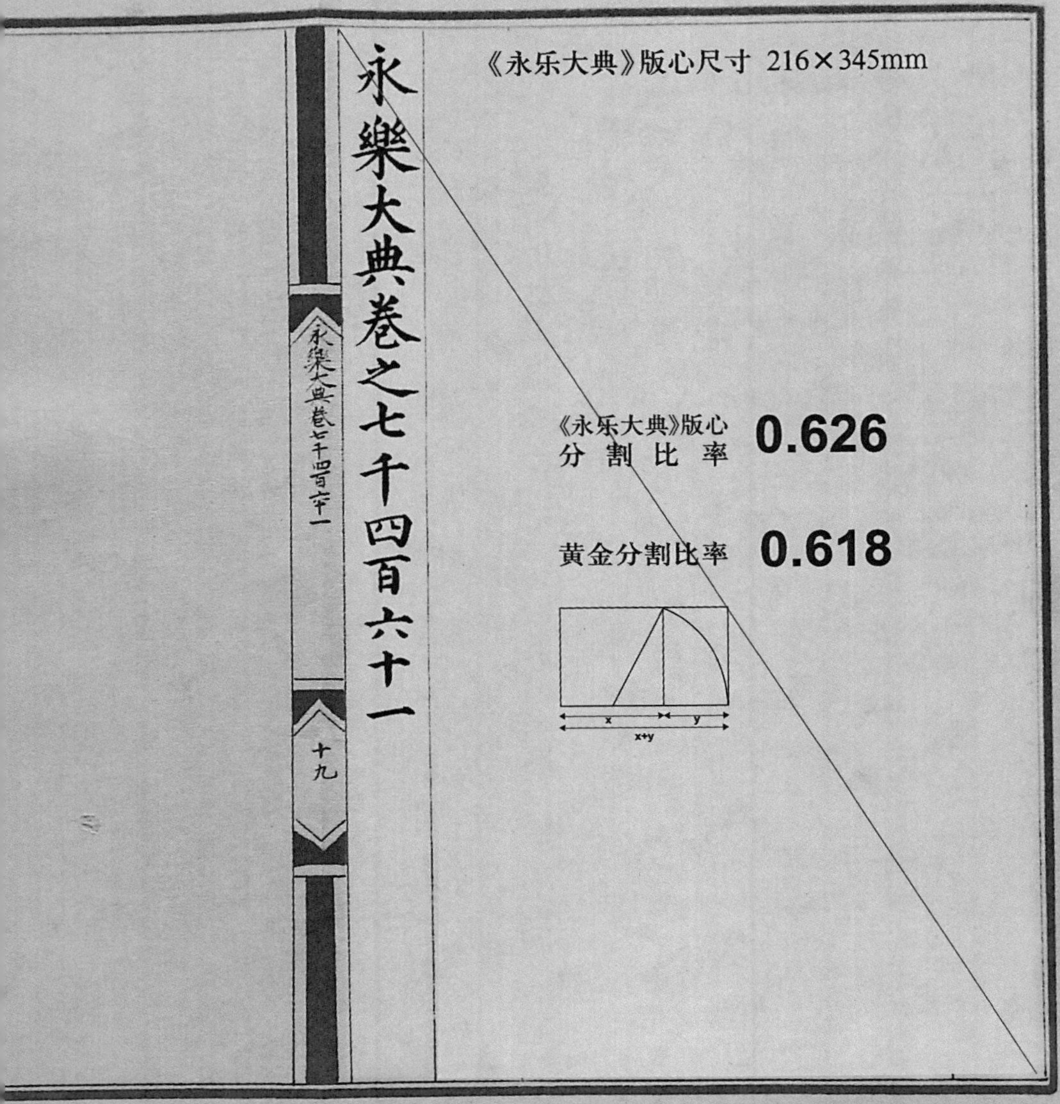

永樂大典卷之七千四百六十一
永樂大典卷七千四百六十一
十九
《永乐大典》版心尺寸 216×345mm
《永乐大典》版心
分割比率 0.626
黄金分割比率 0.618
x
y
x+y

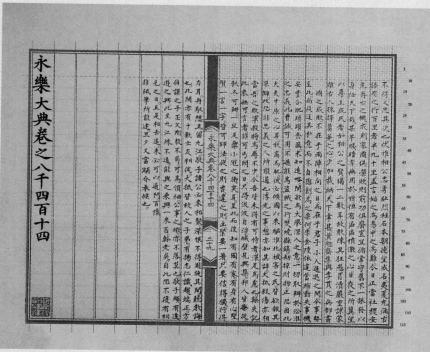

Thoughts on Grid Design

I learned grid design from Kohei Sugiura in the late 1980s. He fundamentally changed me who had been used to just "decorating" books in China. That event truly invited me to the book design industry. Grid design was not only a design method but a way of thinking and an attitude to book design. He said to me, "The word design here means a beautiful process that controls order, while the grid indicating a way that facilitates an efficient and clear understanding of information and makes it stay long. They are a key to the logical solution of a problem. They are flexible and dialectical and can convey two- and three-dimensional information, and even time. Grid design is a must for any graphic designer. Simply, you cannot go without it."

We might raise a question, "Can the grid system comprised of Western letters apply to Chinese characters?" The golden ratio in the use of the typography of the West is meticulous, rational and standardized. This means that the system is established by universal meanings, led by scientific logics. In the meantime, the ancient Chinese phonology has specific rules and built a structural skeleton of language. Chinese characters largely formed in square shapes go better with vector arrangement, so we needed our own grid system. Research of Kohei Sugiura reveals the possibility of introducing the grid system of Western letters to China, supported by relevant cases. Thanks to the principle shown in his book, Chinese book designers could go successful in making the Chinese character typography design. They could learn a design consciousness which is in full control of the beauty of order. They also could imagine their own aesthetic route and practice ways of designing rules in oriental texts.

The era requests us to face, cold-heartedly, countless problems and shortcomings that we have today. Book design should not be a mode that just repeats itself in changing environments, but be a kind of cultural activities. Book designers are not the printing machine that mass-produces identical products. They have to roll up their sleeve and deal with cultural aspects, including texts and design, thought and emotion, and idea and ideology. Book design is not the subject of one-time pleasure. It should keep simplicity long so that readers are moved by it. Put it simply, it must help people enjoy a cultural life. A good book ensures a daily reading. It should play as energy storage that is needed to sustain the vitality of reading.

Books are of course a product, a cultural product. Those in the book industry have to consider profits when planning a book, by laying out an attractive design concept and sales increasing idea. A beautiful book has texts that enable a reader to truly know the charm of reading. It also allows unique readings. In addition, it has logics that consider all things and attract reader's participation in humanities in a careful way. For their part, publishers should not be solely driven by profits. They also have to work very hard on creative activities. By doing so, they will live up to the expectation of readers in the digital age. They have to use this attitude even when checking the possibility of a book.

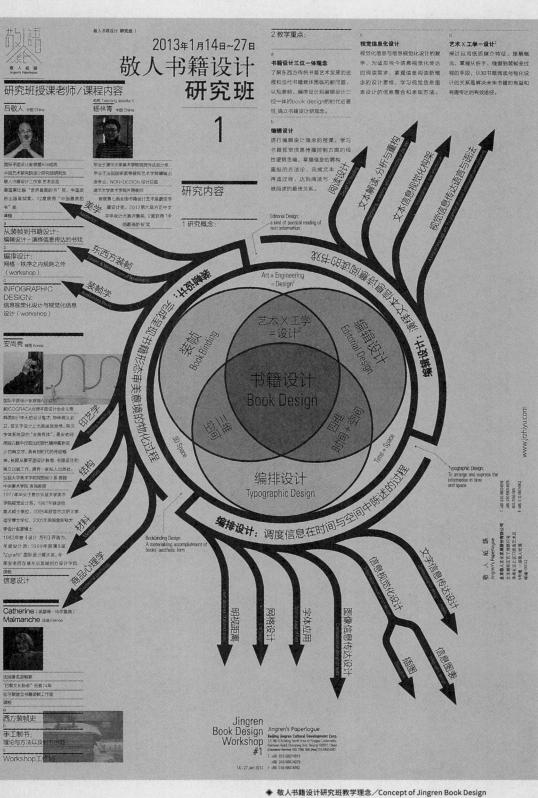

◆ 敬人书籍设计研究班教学理念／Concept of Jingren Book Design Workshop

社会化设计教育的实验性探索：敬人书籍设计研究班

电子载体改变了纸面书作为传递信息的唯一途径的局面，然而千百年形成的书籍信息编制结构、物性造型特征和独特的时空阅读形态却无法被改变。尽管数码技术使出浑身解数进行仿效，仍无法回归"书之为器"自然之妙有的存在感。未来，书籍将在更广泛的领域释放其独特的能量，使更多的设计工作者抱有极大的热情和发挥无穷的创意，让受众享受到书籍和纸文化的魅力。面对挑战，从事设计、教学、出版、印艺工作的人士需要调整和提高自己的学识。

"敬人书籍设计研究班"由享誉国内外的书籍设计演出者共同授课，致力于当代中国书籍设计艺术的理论研究和创作实践，关注国际信息载体发展潮流，成为该领域新设计论的探索者与实施者。研究班通过授课、学术讲座、Workshop、研讨互动、手工书工作坊，国外考察等多种教学方式，开拓书籍设计视野，更新专业设计理念，丰富设计语法与语言，提升实际创作能力，探索新的设计教学方法，重新认知书籍未来的价值和方向。

Explorative Study on Socialized Design Education: Jingren Book Design Workshop

E-Books became an addition to the information delivery system which was solely occupied by paper-based books in the past. The new medium might imitate the aspects of regular books shaped over the last millennium, including, the structure of editing and making information, features of figures, unique time frame and spatial reading. Yet, it would not replace the natural and subtle meanings that paper books give. In the future, books will uniquely relax their muscles in broader areas. Numerous designers will spread the attractiveness of books and paper, led by great passion and boundless creativity. In the face of challenge, those in charge of design, education, publication and printing need to elevate their knowledge.
At the Jingren Book Design Workshop, book-designing artists from home and abroad deliver joint lectures. They are making efforts for the study on the theory of contemporary Chinese book design art and the practice of creative activities. They perform research and execution in this field, as they have kept their eyes on the advancement of information media of the world. The workshop provides various curricula that include lecture, academic forum, training programs, research discussion, hand-made books and overseas field trips. The facility also expands perspectives of book design; upgrades professional design idea; and develops diverse design languages. Through these endeavors, they are strengthening the practical creation capability among students. Adding to that, it explores new design teaching methods as well as values and direction of books in the future.

◆ 敬人书籍设计研究班手册
The Guide of Jingren Book Design Workshop.

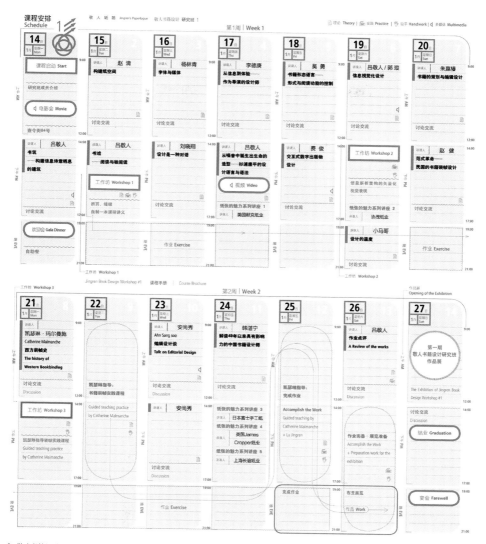

◆ 敬人书籍设计研究班课程表
  Course schedule of Jingren Book Design Workshop

1

2

◆ 1—4
荷兰设计家伊玛·布、日本设计家铃木一志及法国装帧师凯瑟琳先后在敬人书籍设计研究班教学
Dutch designer Irma Boom, Japan designer Suzuki Hitoshi and French Book-Binder Catherine in Jingren Book Design Workshop

◆ 瑞士书籍设计家罗兰特和日本书籍设计家祖父江慎在第七期敬人书籍设计研究班教学／Swiss designer Roland Stieger, Japan designer Sobue Shin in The 7th Jingren Book Design Workshop

Lessons from My Parents / 1947–

韬规家训

感恩父母把我带到这个世间，他们的人生路途在多变的世道中坎坎坷坷，却从未放弃对五个孩子的严厉的传统家教，『子不学、非所宜；人不学，不知义；幼不学，老何为；玉不琢，不成器』，这些话成了励志以工业救国的父亲的口头禅。一生行医，乐善好施，信仰基督，温和善良的母亲，却在淞沪抗战中冒着日寇的狂轰滥炸，勇敢担当前线救护大队长，奋勇抢救受伤的抗日官兵。他们独立的人格和思想修为影响了我的一生。父亲九十六岁时，以我名字中的『敬』字专门为我写了一纸榜书：

敬事以信、敬业以诚、

敬学以新、敬民以亲。

要我牢记行事、务业、求学、做人的道理，遵父母之教诲，这四句话已成为我余生为之努力的座右铭。

I am thankful to my parents for bringing me into this world. Though they had to undergo extreme hardships in their life, they went very rigid in educating their children. My father, who was committed to advancing China through industrialization, habitually cited phrases of Sanzijing, a book for children's education. He repeated, "It is not right if children do not study. People will not know righteousness, if they do not study. What can you do when you grow up, if you do not study when you are young. Jade will not be used for precious wares, if not properly processed." My gentle and nice mother who had Christian faith was a doctor and provided good services to people. When Japanese Imperialist Army invaded Shanghai, she became a battalion commander and gave medical services to the wounded. Independent personality and ideas of my parents were influential to me. When my father turned 96, he made for me a framed picture, by putting jing (敬), one of three characters of my name on it.

To deal with matters with trustworthiness (敬事以信)
To manage business with sincerity (敬业以诚)
To respect learning with newness (敬学以新)
To get along with others with intimacy (敬民以亲)

I respect my parents for their ideas and lessons that are about behaviors, work, study and moral duty. I will do my best to live by the message of my father through the rest of my life.

◆ 96岁的父亲题写家训
96 year- old father writing family instruction

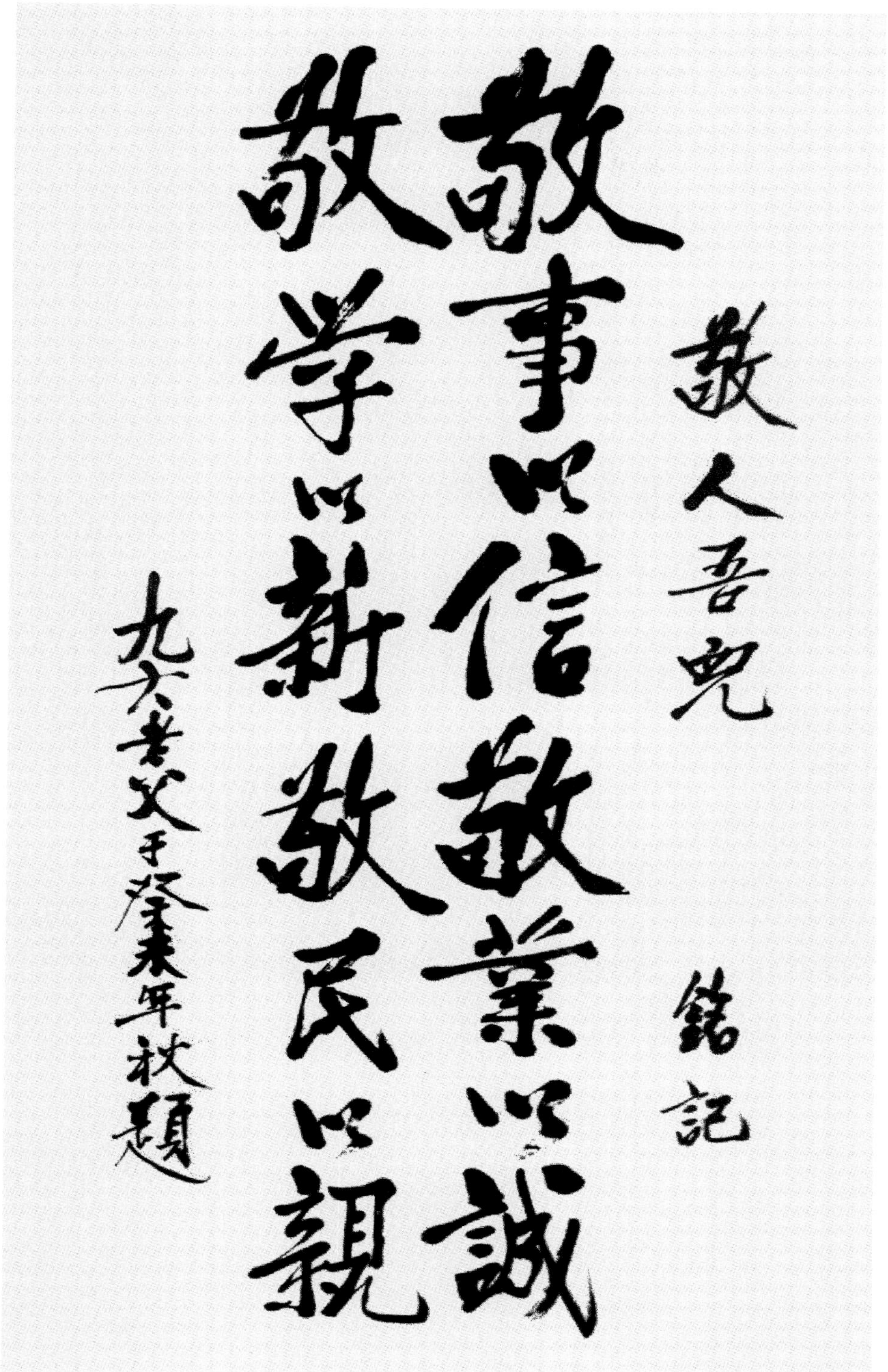

敬人吾兑

敬事以信 敬业以诚 敬学以新 敬民以亲

铭记

九六岁父于烽秦年秋题

◆ 敬事以信、敬业以诚、敬学以新、敬民以亲
To deal with matters with trustworthiness, To manage business with sincerity,
To respect learning with newness, To get along with others with intimacy

◆ 13岁学画
Learning Chinese Painting when 13 years old

自述

本名吕敬人,字守良,1947生于年上海,属相:猪也。愚,命薄,且经常成为人们舌尖上的谈资,予人类有点贡献,也算没白活。

自幼学画,"文化大革命"期间失学,赴北疆农场劳作十年,"文化大革命"结束,遂入出版社,自学专业不殆,甚感理念不足。两渡东瀛求学,幸得恩师杉浦康平点拨,豁然开朗,返国励志推新,论理授教践行,尽绵薄之力。

我爱收集同族玩具,有的笑容可掬,有的温柔敦厚,有的大智若愚,众猪相似乎有些我的影子。今年已到古稀之年,经历不少,涉世却浅,猪性难移,执迷劳碌。猪虽智商不高,又无远大梦想,但安然处事,静观自得,也能活出一番滋味来。

感恩家训:明志行善,敬人人敬。感恩有贤妻孝子的幸福家庭,有感与儿子合作的默契只需一个语气或眼神。感谢那么多年来给予我厚爱的那么多国内外同道好友,因为你们,使我有勇气面对明天。

2017

Self-description

Name: Lu Jingren (childhood name: Suryang (守良) Year of Birth: 1947 in Chinese zodiac belt of the Pig. The reading is that pigs are foolish, short-living and subject of people's talks. Those who have the belt are said to live a meaningful life as they will make contribution to humanity, though it might be minor.

I learned how to draw pictures in my childhood. But I lost a chance to continue with it during the Cultural Revolution. I was sent to labor in farms in the north for 10 years. When the political event came to an end, I found a job at a publishing company. Having felt that self-taught knowledge is not sufficient, I made my way to Japan. In the country, I met Kohei Sugiura and learned lifetime lessons from him. Since I returned to China, I have made efforts to pursue a novelty in my work and have nurtured my students.

◆ 幼年五兄弟
5 brothers in childhood

I like to collect toy miniatures of pigs, "the same species as I am." One of them looks smiling, while another in gentle face. Yet another seems cunning but foolish. Whenever I see them, I feel like they are another "me." Though just shy of 1 year to 70, I am not close to understanding how the world goes. I am simply naïve and sticking to my work. Pigs have neither a good IQ score nor an ambitious plan, but they have their own virtue. They know how to satisfy themselves and evaluate what they already have.

I feel grateful for the domestic instruction of my family. "People Should do good to others. If you want to be respected, the best way is to respect to others." (感恩家训: 明志行善, 敬人人敬)

My wife is wise, and my children are filial. I live with them happily. I work with my son, and we know what each of us is thinking without telling. I am thankful for all of these. I want to give my thanks to my old friends and colleagues from home and abroad for their unwavering supports. I have to say this, "You have me go brave tomorrow."

◆ 父亲与母亲
Father and mother

◆ 全家福
Photography of the whole family

◆ 五兄弟
Five brothers

◆ 20世纪70年代在海南岛体验生活
Went into Hainan Province to draw sketches in 1970s

1

2

◆ 1-2
1968年下放北大荒农场劳动
Jingren was sent to work in the Great Northern Wilderness in 1968

◆ 20世纪70年代画政治宣传画
Painting a political poster in 1970s

◆ 在北大荒农场劳动十年
Worked in the Great Northern Wilderness for ten years

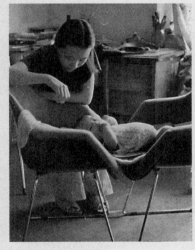

3

4

◆ 3-4
1981年儿子出生
Son was born in 1981

Lessons from My Parents 1947-

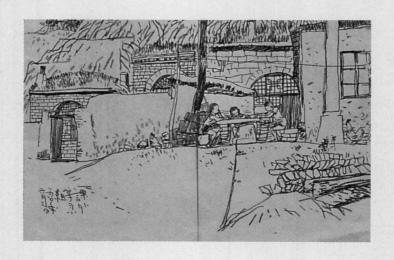

Lessons from My Parents 1947-

◆ 在中国青年出版社工作
Work for China Youth Press

◆ 1989年在杉浦康平设计事务所学习
Began to study in the studio of Kohei Sugiura in 1989

◆ 书籍设计四人展
The exhibition scene of LNWZ: Exhibition of Four Book Designers in 1996

◆ 2003年与杉浦老师在敬人设计工作室录制节目并接受日本电视台的采访
Sugiura Kohei and Lu Jingren was interviewed by a japan TV channel in jingren art design studio

◆ 2012年在奥芬巴赫柯林斯勃博物馆个展中接受法兰克福报采访
Interviewed by Frankfurt Allgemeine Zeitung during the exhibition of Lu JIngren's Book Design Exhibition of Klingspor Museum

◆ 在德国柯林斯勃博物馆举办吕敬人书籍设计展开幕式
Opening ceremony of Dalikat: Lu JIngren's Book Design Exhibition in Klingspor Museum

◆ 2006年在日本东京竹尾举办中国书籍设计展
China Book Design Exhibition in Takeo Tokyo, 2006

◆ 2013年在东莞举办书籍设计五人展
Exhibition of 5 Book Designers in Dongguan, 2013

◆ 与杉浦老师在台北举办书籍
设计展
左1／廖法连（中国香港）
左2／王行恭（中国台湾）
右／马塔（德）
Took part into the book design exhibition with Kohei Sugiura in Taibei
Left 1 / Liao Jielian (HK)
Left 2 / Wang Xinggong (Tai Wan)
Right / Uta Schneider (Germany)

1

◆ 1—2
2006年与杉浦老师、安尚秀、柯蒂特里维迪在新加坡举办亚洲之路四人展
The Way of Asia: Sugiura Kohei, Ahn Sang-soo, Kirti Trevidi and Lu Jingren, Singapore, 2006

2

◆ 与三位弟子吴勇、刘晓翔、韩湛宁在一起
With 3 proteges Wu Yong, Liu Xiaoxiang, Han Zhanning

◆ 2012年在德国莱比锡担任"世界最美的书"评委
Invited to join the jury panel of "Best Book Design from All Over the World International Competition" in Leipzig, 2012

◆ 为中国作家协会副主席、小说家张抗抗设计《赤彤丹朱》
Designed <u>Different Kinds of Red</u> for Zhang Kangkang, Vice Chairman of the Chinese Writers Association

◆ 为中国外交部原部长钱其琛设计《外交十记》
Designed <u>Talk on Diplomacy</u> for Qian Qichen, the former foreign minister of China

◆ 为小说家莫言设计《莫言全集》／Designed <u>the Complete Works of Mo Yan</u> for novelist Mo Yan

◆ 为邓小平之女邓林设计《我的父亲邓小平》写真集
Designed photography Album <u>My Father Deng Xiaoping</u> for Deng Lin. Deng Lin is the daughter of Deng Xiaoping, the former leader of China

◆ 为日本著名建筑家团纪彦之父团伊玖磨设计《烟斗随笔》，与日本著名电影演员栗原小卷一起出席首发式
Designed <u>Essay of Pipe</u> for Musician Ikuma Dan. He is the father of Japanese architecture Norihiko Dan

◆ 为著名京剧表演艺术家梅兰芳之子梅葆玖设计《梅兰芳全传》
Designed the <u>Biography of Mei Lanfang</u> for Mei Baojiu, the son of Mei Lanfang

◆ 为日本著名电影演员中野良子设计写真集《中野良子》
Designed the Personal Album for Japanese actress Ryoko Nakano

◆ 为冯其庸设计《红楼梦——冯其庸纂校》
Designed <u>A Dream of Red Mansions - Proofing and annotation by Feng Qiyong</u> for Feng Qiyong

123　　S ⇨ 315 × 180 mm
　　　　D ⇨ 2006

老父百岁
自出版
为父亲一百岁生日编辑设计
老父百岁一书

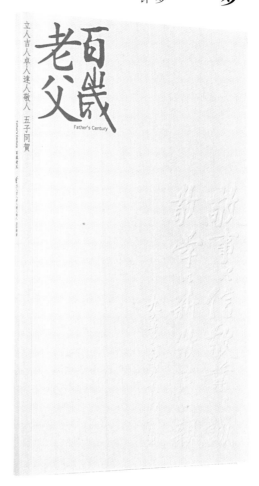

My Father Turning 100

Father's Century is a self-publishing book. It was a good way to celebrate father's 100th birthday.

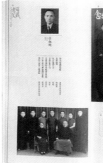

Lessons from My Parents 1947–

远去的旭光

自出版

为纪念赴北大荒四十周年，做旭光农场上海知青回忆录

S ⇨ 240 × 166 mm
D ⇨ 2008

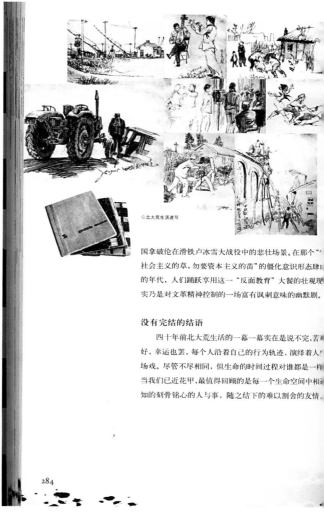

Gleaming from a Distance

Gleaming from a Distance is a self-publishing book. It is for the 40 years memorial of the Movement of "going to mountain areas and the countryside" of 1960s.

情、恩情。其中包含1968年的夏秋之交,北上列车内每个上海知青的希望、理想和经历的种种遭遇,留下了那个时空北大荒人给予我们的宽容、友爱和关照,留下了每个知青相互之间不图回报的纯真,也留下了曾同乘一趟列车北上却永远不能回的消逝的年轻生命。

旭光前面那条南来北往的铁道,寄托着多少人的奢望与梦想,1分钟的停车瞬间截下了多少眼泪与欢笑。每日清澈透底的白菜酱油汤只够铲半垄地的能量,寒冬缺煤戴着棉帽睡觉,早晨起来时发现帽子已被冰霜冻在墙上拔不下来……但大家仍在笑,因为那时还年轻。

朴可爱的大勺子,我初恋"鸡毛"的专职信使王维全

旭光人来自祖国东南西北,历经煎熬几上几下老转业功臣的张场长、呼风唤雨上下逢缘的汽车队程队长、干练豪爽敢说敢当的四连张连长、吹牛拍马扔手榴弹害人的大瞎刘管理股长、心灵手巧的捕狍猎手电影队长贾师傅、多嘴饶舌好管闲事的烧炉工老周头、憨厚淳朴善良能干的文艺宣传队员、当地小伙儿大勺子、一道同甘共苦的下乡荒友们、为追求艺术理想而废寝忘食的北大荒画友们,还有团机关宿舍土屋的大炕,门前那块栽满向日葵的小小瓜蔬地的温馨回忆……旭光的世态百相怎能不让我时时想起,那些人间真情更令我久久难忘。

毋须感谢让我们背负十字架的年代,但也可庆幸自己经受过非常态年代的砺练,并结识了许多同道挚友。我们失去了很多,但也得到了些许。四十年后跨入六十花甲的我们各在东西,有的已进天国。今天,有幸还健康活着的旭光人通过文集重新聚合在一起,因为我们对友情和生命表示尊重,并相信这些文字能感动他人。

2008年正月十五 北京竹溪园

125   S ⇨ 185 × 125 mm
              D ⇨ 2005

自出版 | 我属猪

**I was born in the
year of the Pig**

Born in the Year of Pig is
a self-publishing book

Lessons from My Parents 1947–

318
319

Pupils Walking with Me

同
道
弟
子
说

我想快乐的事莫过于人生多有知己，古话说『物以类聚，人以群分』，有相同的趣向爱好，有近似的为人秉性，有共同的价值判断，谓之志同道合者。十位弟子与我有年龄差别，经历出身各不相同，入道也有先后快慢，有受我影响者，也有我直接教授的学生。看着他们学业上执着探寻，专业上独有建树，事业上成绩斐然；在几十年求道的路上，我与他们亦师亦友，教学同行。他们当中有世界平面设计联盟的成员，有多次国际大奖的获得者，有设计领域中的翘楚，有设计教育的导师，有行业中的新生力量。看着他们的点滴成就，有情不自禁的洋洋得意，并自豪于青出于蓝而胜于蓝。他们在与我交往的经历中留下了以下的叙说。

It could not be more joyful than to have a bigger company in life. There is a saying, "Birds of a feather flock together." The meaning, as you know well, is that those who have same hobby, personality, values and thoughts get together. My 10 pupils and I are different in age, career, background and the timing of entry in this industry. Some of them are influenced by me, while others are my former students. I saw them studying hard; achieving wonderful results in their specific field; and going successful in their business. For the last decades, we have been walking together. Some of them are members of the Association for Graphic Information; a winner of diverse international competitions; an outstanding talent in book design; a design professor; and a rookie spotlighted by the industry. They are better than me, and their development and accomplishments make me really happy.

吴勇 WU YONG

国际平面设计联盟（AGI）成员，AGI 中国分会副主席，中国美术家协会平面设计艺术委员会委员，长江艺术与设计学院平面设计系主任、教授，硕士生导师。
曾获香港设计师协会金奖国家图书奖，首届香港国际海报三年展金奖，全国书籍装帧艺术展览金奖、银奖，首届"中国元素"银奖，"中国最美的书"奖，"GDC07 展"海报类银奖，东京 TDC 奖优秀奖。

Member of AGI, Vice president of AGI China, Member of Graphic Design Committee, China Artists Association Head of design department being a professor and supervisor of graduate programs of Yangtze Arts and Design College, Shantou University.

Gold Prize at the HKDA, Prize at the National Book Awards, Gold Prize at the Hong Kong International Poster Triennial, Gold and Silver Prize at the National Book Design Exhibition, Silver Prize at the China Elements, Most Beautiful Book in China in 2009, Silver Prize in Poster at the GDC 07 Exhibition, Good Prize at the TDC Award, and others.

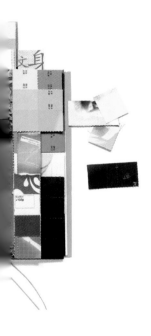
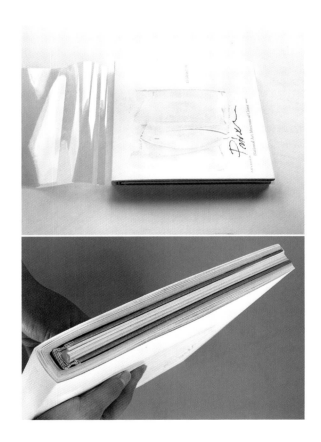

老吕的星座 | 吴勇

初识

我与老吕相识于20世纪80年代，那会儿我还是个初生牛犊，刚从象牙塔里溜达出来，不谙世事，浑身带刺，满眼不屑，大脑袋里装的全是理想。而那时国家仍处在计划经济末期，大学生都必须服从国家分配，来自武汉的我则幸运地留京并被分配到中国青年出版社美术编辑室，而老吕正是这个部门的副主任。

之所以称吕老师为"老吕"，是源于当时美编室的融洽氛围。老先生们都称他作小吕，年轻的我们便顺其自然称其为老吕了。那时的我头发尚存，一头拢于后脑的长长黑发，成天在那帮老先生们面前甩来甩去，晃得他们眼花花的头疼，就没了好印象，便想着在我白嫩嫩的身上挑刺，总觉得我应该有问题，所以若有人基于私利，捕风捉影地向老先生们打小报告，他们一定宁信其有。

但老吕信任我，他在我浮躁的外表下看到一颗宁静而向上的心，批评也总以半玩笑的方式进行，既转达了老先生们的不满，也保护了我的积极性。所以，美编室成了阻挡是非的"避风港"，能让我在这里用十年的时间慢慢成长：窗外，政治风向此起彼伏，人事斗争惊涛骇浪；室内，滋养我的是老吕坚持的专业用心和他为美编室内部订阅的《Graphics》《IDEA》《fonte》等国外专业杂志，还有从国外带回来的各种资讯。

灯塔

我总觉得老吕就像能照亮别人的一座孤独的灯塔，不畏世间的惊涛骇浪，永远孤傲地挺立于忽隐忽现的险象中，以不屈不挠的坚定感给人以航向。总觉得他的这种坚持，冥冥之中来自某种神秘力量，一定是具有某种特质！

也许老吕并不熟悉与他同星座的日本昭和时代的大诗人宫泽贤治，但这却是一位与他在另一个时空平行的人，彼此折射着许多的相似。

宫泽贤治是一个既简单又复杂的人，出身望族却有一颗理想而悲悯的心。为改善农民生活的贫困状况，他创办农会，指导科学种田，还辞去公职，引导农民学会欣赏音乐并鼓励他们演出戏剧，为他们的孩子讲述童话故事等，凡此种种无私的奉献都是为了实现心中的"理想乡"。

出生于上海"非无产阶级"家庭的老吕，从小受到"仁义礼智信温良恭俭让"的传统家教，为人处事待人接物都是温文尔雅、低声细语的。从日本学习归国后，为了心中的理想国，他抛头露面以讲座的方式传播着当时稀有的设计思想资源。其实他特别不善于应酬，但讲专业可以滔滔不绝，对付官场等其他的公开场合则显得特别生疏与生硬，其实很是难为他的。大约90年代以来，他便以数百场讲座的纪录几乎踏遍了国内所有重要的出版组织和机构，传播新理念，许多的美编、出版人及大学师生都听过他的演讲，人数不下万人。这是业内无人能超越的纪录，这是一种理想的宣泄和真挚而无私的奉献。老吕从演讲初始的生疏羞涩，至云游书籍审美的形而上和设计方法论，始终谦逊恭敬地面对每一场的演讲。人们无不被他娓娓道来的书理、循循善诱的风格所折服，许多人听几遍也不觉枯燥。我相信当那些捕风捉影、人云亦云的不以为然者，在聆听了老吕的现场演讲后，也会被他带到"沟里"，鼓掌钦佩于他的学识。这都源于他千百次演讲背后追求专业的进取心和精了又精的"认真劲"。

痴人

宫泽贤治的所为虽然和其富商之子的身份不符，但他却因为信仰而乐在其中。他有着多重又丰富的身份。他既是诗人、作家，又是教师、农业改革者、宗教家，但他更是一个自始至终怀抱着梦想、单纯又真挚的人。

他终其一生，不安宿命，犹如一匹奔腾的骏马，为自己所念所想忘情奔驰。

他那超越常理的人生选择不为亲友及世人所理解，无人赞扬，更谈不上歌功颂德。他创作了千余首诗歌，只有其中一篇童话《渡过雪原》得过稿费。后来，人们才发现其在孤苦岁月中写下的那些零乱书稿，竟是如此的五彩缤纷！于是在他身后那些作品终于得到了世人的瞩目，被译成14种文字出版，还被收入教科书，甚至被改编成电影、电视，以他作品里面主人公命名的食品、商品不胜枚举。很多日本国民对诺贝尔文学奖获得者川端康成不甚了解，但却非常了解平凡而不凡的宫泽贤治。

与宫泽贤治同为处女座的老吕，是设计大家，是画家，是作家，是教授，是公益活动家……他的多重身份只为一个目的：以更多人的力量改变中国人的阅读品质。

以人口的基数来看，中国本应是个阅读大国，但实际上却落后于很多国家。中国图书出版量世界数一数二，每年出版50万种读物，册数近百亿。但中国的图书馆资源并不丰富，藏书的流通率很低，人口过10亿，但全国图书馆持证读者只有500万方，占总人口数不足0.5%。国人阅读量严重偏低，人均年阅读量不超过五本，一本200页的书，中国人愿付的平均价格为13.67元，仅约为一杯星巴克冰拿铁价格的一半。在为了效益至上而只图书衣打扮的装帧领域，陈旧观念淤泥翻滚，浑浊不堪。

几十年来却依然有一股清泉在默默流淌着，沁过许多设计人的心脾。今天，这股清泉不只是老吕一个人在孤独地流淌，它汇聚了众多的志同道合者在涓涓长流。近些年，我们的出版物品质在上升，除了众多书籍设计获"最美的书"奖，在书店里也出现了许多高品质的读物，"养成"了一批高品位的读者，这是可喜的，这是国民素质得到启蒙的直接表现，当然这里面有老吕痴人执着的信念坚持和他身体力行的不断努力与呼吁。

包容

据说处女座的人是知性的，做事一丝不苟，有旺盛的批判精神（那是因为他们总希望世事能符合他们的主观标准），是完美主义者，极度厌恶虚伪与不正当的事。同时，都持有一颗赤子之心，充满着对过去的回忆及对未来的梦想。通常他们又很实际，并非常理性，强调完整性，不喜欢半途而废；对任何事都有一套详细的规划，然后一步步地实施直至完全掌握；他们对自己的要求很严格，从不妥协、让步，是个优秀的"工作狂"。这些完全是老吕的写照！

出版界熟知的历届"全国书籍设计展""大学生设计展"和"疾风迅雷——杉浦康平半个世纪杂志设计展""书籍设计四十人邀请展""书籍设计四人展"等，无不有着老吕艰辛的身影。他除了筹措资金、策划流程，更是借此团结一切热爱书籍设计的人士，构建一个爱书的理想国。这其中，不分成绩优劣，无论地域远近，对一些天资不够的努力者也不嫌弃，他都一视同仁给予力所能及的帮助。于是，有的成为了卓有成就者，有的成为书籍设计专才，有的成为教育界广受欢迎的优秀教师，当然也有的原地踏步，有的则改弦易辙不再坚持……所以有的人会认为老吕没原则识人不清，其实他们狭隘了，老吕不是在弄精英帮派，他是在努力营造好的造书环境，推动中国书籍设计事业的整体前行。

1996年，老吕筹办了"书籍设计四人展"，参展的四位当时均为出版社美术编辑，而在当时一个设计类的非官方展还属异端，老吕就是希望排除论资排辈的传统意识，利用民间力量发出学术的声音！在新开业的三联书店二楼隆重开展的"书籍设计四人展"，引起出版界极大轰动，无论展览表述方式还是观念都是具前沿性、宣言性的。同时在出版的《书籍设计四人说》中设计了一个独特的视觉符号，即各取了每个人姓氏中的一个偏旁部首，构造了一个新汉字，念"sì"。书中特别提出"书籍设计"的理念，质疑了"书衣""嫁妆""书装设计"的狭隘理念，把书籍设计的工作提升至与作者同步的层面，从意识上具有与作家并行的"第二作者"的专业担当，双方的关系是出版物形态构成的不可或缺的共同体，这超越了出版社美编原来规制的职责范围，对体制内的设计思维是一次极大冲击，可谓是前所未有的。在当时的四位参与者中，老吕已是从东瀛杉浦康平老师事务所学成归国的业界翘楚，另一位则是做书的前辈，颇有声望，而还处于懵懂做书阶段的我，当时有一股敢胡思乱想的冲劲，有幸得到老吕的认可和赏识，被纳入其中，并委以重任，做展览形象及作品集的执行设计。今天回顾，老吕当时希望打破业内论资排辈的惯习，关注专业，提携新人，建立真正健康的学术交流新风，这样的例子不计其数。四人展已过去 20 年了，一拨又一拨的人被他的精神感染，一批又一批后来者经他的包容和鼓励，成了中国当代书籍设计的中坚力量。

师德

另一位具普世胸怀的伟人是特蕾莎修女，她与宫泽贤治同样都是跟老吕同月且晚一日生辰，当然他们年长老吕许多，但都是处女座的贤德之人。

特蕾莎修女把自己的一切都奉献给了穷人、孤儿、病患者、孤独者、流浪者和临终者。她从 12 岁起直到去世，用了 75 年时间，只为受苦受难的人而奔波。特蕾莎修女以博爱默默地关注、帮助着苦难中的人们，使他们感受到尊重和仁爱。心如静湖的特蕾莎修女没有高深的哲理，仅用诚恳、服务、施爱，来医治人类邪恶的病灶：自私、贪婪、剥削、享受、冷漠、残暴等，为引领人们通向社会的正义和世界的和平，开辟了一条光明的通路。

但不是所有人对特蕾莎修女的看法都是正面的。生于加尔各答、现居伦敦的批评家查特基 Aroup Chatterjee 宣称"在加尔各答她从来就不是一个重要的存在"。他指责特蕾莎造成了外界对他的家乡的负面印象，并认为特蕾莎修女时常反对印度民族主义的做法在印度政界造成了不必要的隔阂与摩擦。即便是一个备受争议的人，也湮灭不了她积极的一面，特蕾莎仍是一位得到公认的伟大的凡人！

老吕提出从装帧到书籍设计观念转换的整体设计概念引发的争议和误解也是不断的，纯专业的观点被上纲到政治，但不畏威胁诬陷的他仍坚定地行走在自己的路上，坚守中国设计必须与时俱进的信念，扎扎实实做好推动中国书籍设计进步的每一件事。

比如，他为书籍设计领域做的每一个全国性大赛，为设计教育主持的每一届大学生展，为行业主编出版的《书籍设计》杂志，都是用个人时间，全身心投入，无偿付出。更多人以为老吕从清华退休后办书籍设计研究班是为赚钱，其实他分文不取，还自掏腰包去国外邀请国际大师来国内授课，每次课程汇集各路翘楚组成十几人的教师团队，耗资巨大。他说："同学付出的学费都要花在他们身上。"这就是他的办学理想，他希望以一己之力提供多样资源，让更多的人收益，并传递有价值的设计意识与审美思考，靠更多的同行改善我们尚不乐观的现状。

当然，听过他的课的人无不为其魅力所感染，获益良多自不必说，更多的是他那循循善诱的教学方式让人心生愉悦，课程总是那般吸引人。这些当然与他的为人师表的教学态度和专业精神相关，无论在哪个讲台，他始终如一：清华本校的、其他外校的、工作坊的、学习班的，哪个他都恪守职责、尽心尽力。

有一年，我请百忙中的老吕去汕大上一个月的书籍设计课程，他爽快地答应了。其实我当时是想让忙得没白天没黑夜的老吕借机休整一下的，结果却大大超越了我的美好想象。去机场接机，只见老吕一瘸一拐地被空姐扶出来，他的痛风病犯了！我问："为什么不告诉我，可以调课甚至取消的呀！"他老人家居然还笑眯眯说道："你是在让我犯错误吗？答应的事是不能违约的！"我的天呐！接下来，住在酒店连吃饭都下不了床的老吕，坚持要我弄辆轮椅车，由四个男同学抬上抬下阶梯，坚持上完全部课时，包括一场讲座！一切照旧！一个不差，一个不少地扛了下来！

上课时，从他不时抽搐的脸上可以看出，老吕是强忍着剧痛在坚持，但整个课程丝毫不受影响，他依然是声情并茂地讲述与讲解，从理论的分析到具体的辅导，一丝不苟。两个班的同学以及众多旁听的本科生、研究生们也为老吕的精神所折服，仔细听讲、认真笔记、发散思考、积极提问、认真作业……最后结课的课程展非常有成效，这应该是一次令汕大师生难以忘怀的高师德与高水准的课程了。

因为他对自我及对身边人的高要求，让我们时时不敢苟且自己。在汕大，也有人认定我们这帮非体制内的"空降兵"早晚会撤退；直到这么多年坚持下来，人们又在质疑我们不撤一定是有什么好处，或者是在等待什么好处。我想，其实我更能理解老吕的坚持了，这种坚持让自己的内心充实、愉悦、不知疲倦，有一种真实的存在感。虽然我不是处女座的，但我有很多朋友都是这个星座，他们让我觉得这个无序的世界还是有要求的、有标准的、有美好，常常可以激励到我、警世于我、推动着我！

老吕不是伟人但却是值得尊重的人，这篇让老吕看后鸡皮疙瘩掉一地的拙文，虽酸却很真实。这种感受只有与他经历过许久才会感同身受，我便是那个跟随老吕一路走来，由飘逸长发渐渐头顶稀秃的同道之人，每每在理想破灭之际，总有座灯塔如警灯般闪烁不停，使我振作而不颓靡，这是一股可以一直支撑我与现实不懈抗争的力量。

人生七十古来稀，老吕老了，我也老了。他多年的不懈奋斗、隐忍抗争，让他有了更多的坚持。我们可以质疑他许多的坚持是否过时了，是否不合时宜了，但他就是这样一路坚持过来的，就是这样地影响了众多的人，就是这样创造了出版界的一片生机。一位出版前辈说得好，老吕正值"盛年"，理当继续发光，老吕老父的百岁基因，不仅可以让他肉身长寿，更是可以期许其思想的青春不老，因为他总爱跟年轻人混在一起，除了辅助他们成长，也从中获取年轻感，所以老吕是不喜欢过生日的，他不服老。

此文写在老吕七十大寿之际，耳顺的我祝他古稀今不稀，长命百岁长！因为我需要航标……

"度娘"说：

处女座男人在女人面前，大都给人以有思想、有智慧的印象；在男人面前，往往给人以温文儒雅之感。

与处女座男人共事，轻松愉快。他用持续不断的自我克制或批评精神来自卫。把遵守习俗作为他行动的准则，一成不变的生活是他理想的天堂。这是个既认真又有强烈责任心的人，一丝不苟地做好本职工作是他的座右铭。所以，可以充分地信赖他，依靠他。一般来说，这种人不喜欢到独立的事业中去发挥才智，而愿意在专家、教授或领导者的身边做助手工作，因此，他所得到的利益与他所付出的辛苦一般是不相称的。然而尽心尽责地完成本职工作是他的信条，也是他引以为豪的最大心愿。

用四个字来形容处女座男人——好好先生。再用六个字来归纳性格——有贼心，无贼胆。大部分处女男都类似金牛男，对感情比较内敛，不轻易地表露出来，这个大概和他们有一颗敏感而害羞的心有关。

他们都是很注重礼貌和仪态的优雅男子。

这次，"度娘"的话还是对的，因为老吕就是如此这般的一个人。

刘晓翔 LIU XIAOXIANG

国际平面设计联盟（AGI）会员，中国出版协会装帧艺术工作委员会常委，高等教育出版社编审，刘晓翔工作室艺术总监。
曾3次获得德国莱比锡"世界最美的书"奖，14次获得"中国最美的书"奖，2013韩国坡州出版奖·书籍设计奖，第3届中国出版政府奖·装帧设计奖。

Member of AGI, Executive director of Book Design Committee of China's Press Association, Editor of China's Higher Education Press, Art director of Liu Xiaoxiang Studio.Three awards from the "Best Books from all over the World" in Leipzig, Germany and 14 prizes from the Most Beautiful Book in China. Paju Book Award in Korea in 2013 and Book Design Award of China's Press Government Award.

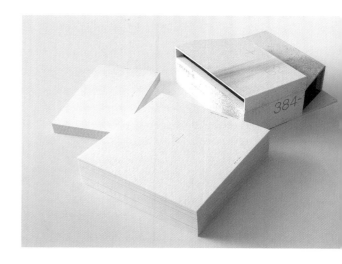

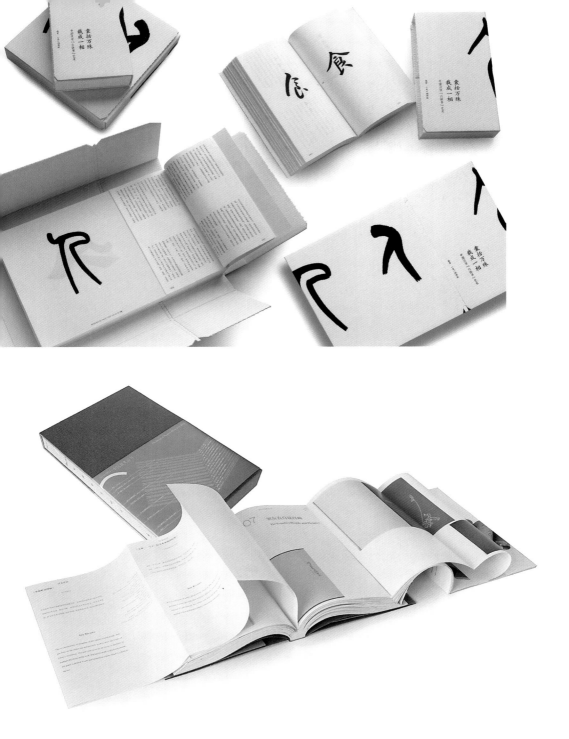

恩师 | 刘晓翔

从不懂设计到略窥门径,我走过了漫长的求师问学之路。

在我上大学的20世纪80年代,设计还不是一个学科,我对于设计的认识基本停留在装饰、装潢和画花边的阶段。油画高大上,国画次之,考不上绘画的才学工艺美术。从心里讲我那时是瞧不起装潢的,也从未想到我将来会从事由装潢进化而来的平面设计并且真心热爱它,真是时也命也!

来北京就是为了圆我做一个画家的梦想,为此一心二用了10年之久,甚至先生来到我所在的高等教育出版社讲学我也没去听,因为书籍设计和当时的我似乎没有什么关系。

1997年左右,我的一心二用走到了该做出选择的时候。或许是来自冥冥之中的无形召唤,在我犹豫、彷徨不知所措之时,一个偶然的机会我看到了《书籍设计四人说》。啊!怎么会有这样的设计理念!书籍可以这样设计!能这样表现设计者的想法啊!这么有质感!它为我把紧闭的设计之门推开了一条缝隙,对我产生了莫名的吸引力,逐渐成为我表达态度、传递爱憎、承载快乐与痛苦的平台。

由绘画进入设计很难!它们一个感性一个理性,仅凭绘画的两把刷子和感觉是做不好设计的,不同类别的书籍也绝不是仅凭感性就能驾驭。我一定要为自己找一个老师,这个老师就是我现在的先生!但那时对先生能否收我这个"半路出家"的弟子心里一点谱都没有。

从幼年到少年,我的成长环境严酷苛刻,对于一个"可以教育好的子女",除去看惯了的白眼与如影随形的蔑视,我是不敢有自己的梦想的。妈妈说:"你一定要好好学习,学习好了才能有机会将来当一个工人!"深深的自卑就这样一直伴随着我。

先生对我来说是只能仰视的神,我不敢贸然接近但又渴望得到他的教诲。

2004年的"第6届全国书籍装帧艺术展览",我以一套《中国历代美学文库》整体设计获得社科类金奖,先生在《第6届全国书籍装帧艺术展览优秀作品选》中点评了我的作品,并让我代表获奖者在颁奖大会上发言。这让我欣喜若狂!终于,我有了求师之机会。

其实,先生是温和、体贴、幽默并循循善诱的。一次,我拿着一本书的贴样请他看,看过之后先生问我:"晓翔,你是怎样设计版心的?""凭感觉,怎么好看就怎么设计。""噢,如果能理性地规划版面是不是更好呢?比如,让版心的宽与高能被3或4毫米整除,甚至,开本也能被3或4毫米整除,可以试一试。"就这样,在不让我感到狼狈的情况下,先生教给我书籍设计的基本法则,将我带到了一个丰富、理性而充满五感的书籍设计天地。

我的喜悦和烦恼总愿意说与先生,诙谐与幽默之间它们就被先生化解掉了,所谓得意的时候不忘乎所以,失意之时也绝不气馁!不管为设计出好的书籍与作者、出版人打交道有多么艰难,我都在先生身上看到、体会到他处处为他人着想又不放弃自己原则的忍让与坚持。我的个性刚而直,追求完美却常常在现实面前灰头土脸,真的是"理想很丰满,现实很骨感"吧。每逢这种时候,先生说的"退两步进一步也是进"就会提醒我。我觉得,在我向先生问学的这些年里,先生所赋予我的不仅仅是设计的方法论和技巧,更多的其实是对人、对事、对物的态度,恩师于我形同再造,先生视我如子,我敬先生如父。

2011 — 2013年,我协助先生在多个场所办了多场书籍设计展。办展览可不是一件轻松的工作,从策展到展览布置、展品摆放、展签制作、展览海报设计以及学术讲座,先生都事必躬亲,不放过每一个细节。多次见他累得几乎可以随时倒下,却不愿意听从我们的劝告暂作歇息。那几年,先生屡次因为太劳累导致痛风发作,展览却总是克服各种意想不到的困难而如期举办,即便是痛风发作期间先生也要拄着拐杖与大家一起工作。"见微知著,如果不遵守承诺做到和做好每一件小事,哪里能做什么大事呢?我们做的都是小事,一本书,一个小小的宇宙,能做好就不容易啊!"

先生对弟子关心爱护,看重他们成长和迈出的每一步;多次推心置腹的交谈与教诲,帮助我克服学术上的困惑与设计上的难关。"晓翔,这一段时间你的设计是不是太追求高大上的黑白双色了,给我看一点色彩好吗?"遇到我设计得不到位,先生总是快步跑向他那层层堆叠的高高书柜,拿来我能够参考的书:先生曾经成功处理的设计案例或者相同问题,抑或他所藏的其他设计师的杰作。甚至,听说我还没有杉浦老师的《全宇宙志》,先生先是托佐藤先生在网上查找,看看哪里有最好的;去东京时在痛风尚未痊愈的情况下忍着疼痛,一瘸一拐地到神保町为我买来品相上佳的第一版《全宇宙志》……

在先生的鼓励下,2011年我在雅昌举办了生平第一次讲座,尽管我站在讲台上双腿发抖,先生还是用他温和的目光不断地鼓励我。此后,不论我在哪里讲座或者在"纸语"上课,先生听过与否,他都会非常认真地听并把不足之处告诉我,"讲一次就是自我总结一次,善于总结才能进步"。其实何止是我讲的,不管是弟子还是别人的讲座,哪怕是他曾经听过的,先生都会极认真地倾听,他总是说"我看看你们是不是有新东西了,俺也学学"。

先生最喜欢看到我的设计,有了想法我也愿意先说与先生,请他教诲,做了贴样甚至打出样书也一定要请先生看看。究竟请先生批改过多少次设计到现在真是数不清了,印象深刻的至少有《中国美术史十五讲》的"中国美术发展概况意象表",《恶之花》的色彩运用,《离骚》的文本空间分配,以及《农民中国:江汉平原一个村落26位乡民的口述史》《书籍设计》《2010－2012中国最美的书》……

从编辑设计理念到网格系统再到信息设计,是先生带着我走出封闭的环境与狭窄的视野,直到在恩师鼓励下,2012年我开设了自己的工作室,2015年经恩师提名加入国际平面设计联盟(AGI)。

从"装帧"一词被引进到"中华民国"再到它成为新中国《现代汉语辞典》中的一个名词,"装帧"的概念与内涵一直没有进化和清晰界定。概念不清也就意味着什么都可以往里装,旧瓶新酒也好新瓶旧酒也罢,由"装帧"进化到书籍设计成为最大问题。尤其是总有人觉得凡事越古越好,对于那些百年千年老汤都敢"喝"的"传统"卫道士们来说,不足百年的"装帧"已经是中华文化的代表了,完全不需要与时俱进。

先生几乎以一人之力不顾他人菲薄,带动了一个行业的发展,引领我们在理念上与国外同行齐肩。先生自己的设计更无需多说,他站在时代的前沿并引领了中国书籍设计的新时代,为世界贡献了来自现代东方的设计理念与审美追求。

先生清晰界定了书籍设计的概念与内涵:装帧、编排设计、编辑设计共同构成书籍设计。清晰的概念与内涵为从事书籍设计的平面设计师指明了方向,将设计延伸到以往不曾达到的领域。这绝不是单纯的视觉表现力

◆ 素描／刘晓翔／成品尺寸210×280mm／1989／1995

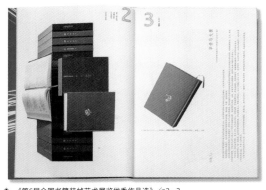

◆《第6届全国书籍装帧艺术展览优秀作品选》／p2－3

扩展或设计的延伸,不是在纸空间之中粉粉刷刷,而是对书籍设计师逻辑思维能力的推进,并由此产生书籍由感性到理性再回归感性的美!

书籍设计使设计师从为文本做嫁衣晋升为令文本视觉化呈现并物化的导演,这是书籍设计概念的一个维度,它促使设计师从文本之中抽象出具有逻辑关系的视觉传达形式,为文本安居页面增添设计师的个性表达,做到了相同文本之书籍"这一本"由表及里地完全不同于"那一本"。在文本可以经由不同载体呈现的今天,个性迥异几乎就是纸本生存的基本前提与未来方向。"留住传统纸面载体阅读温馨的回声。"

书籍设计概念的另一个维度是对于出版人而言,拓展了书籍的可"开发"空间,也引发对于书籍这个传统文本容器之边界的思考,它是对"服务"于文本这个传统理念的解体与重构,将容器与本源容纳物分离,引发出版人对出版物的丰富联想以及对作者、设计师、出版者和读者关系的深入思考。它使点变成了面,平面进化为立体,单通道成为多轨交叉。那么,可以阅读的就不仅是文本,还有编辑理念、构成方式、文本延展与设计本身。

工艺与材质结合传递书籍作为物的五感,让读者体验"阅读的温度",成为书籍设计概念的第三维度。溯往为鉴,工匠与工艺在汉文化传统中一直形而下,书籍设计把工业化制书的工艺与材质提升到新高度,散落在各个环节的珍珠被这一概念连成一体,赋予材料、工艺美学之上的意义并将之视为阅读的延伸。

设计师的文化身份是恩师念兹在兹的,母语构成了汉语语境之下书籍设计的基本脉络。丢掉自己的文化身份认同等于把自己变成水面上的浮萍,"不摹古却饱浸东方品位,不拟洋又焕发时代精神"!无论何种题材与书籍类型,传递汉语与汉字之美是先生一以贯之的美学追求。这也深深地影响了我,成为我设计书籍的理念和目标。面对炫酷的西方,我何尝不被吸引而产生犹豫与彷徨,逢此时,先生的作品与教诲犹如钢琴的调音器一般,帮助我校准音阶,明确努力方向。

……

小马哥 XIAO MAGE

国际平面设计联盟(AGI)成员。获得奖项包括：2017年美国纽约THE ONE SHOW金铅笔奖，2017年美国纽约第96届"国际艺术指导协会"(ADC)银方体奖，2017年中国北京第4届中国出版政府奖（装帧设计奖），2015年美国纽约THE ONE SHOW铜铅笔奖，2014年德国莱比锡"世界最美的书"奖，2011年德国莱比锡"世界最美的书"奖，2010年美国纽约第89届"国际艺术指导协会"(ADC)铜方体奖（两项），2009年中国深圳"GDC平面设计在中国双年展"金奖，2009年中国北京"第7届全国书籍设计展"最佳设计奖（两项），2009年中国香港"香港设计师协会亚洲设计大奖"银奖，2008年美国纽约第87届"国际艺术指导协会"(ADC)银方体奖，2007年中国深圳"GDC平面设计在中国双年展"全场大奖、形象识别类金奖、出版物类金奖，2004年中国北京"第6届全国书籍设计展"金奖，2006—2016年中国上海"中国最美的书"奖（多项）。

Member of AGI. Won Several Awards, Including Gold Pencil from THE ONE SHOW, New York (2017), Silver Cube from The ADC 96th Annual Award, New York (2017), Book Design Award from The Chinese Government Publishing Award, Beijing (2017), Bronze Pencil from THE ONE SHOW, New York (2015), Most Beautiful Book Prize from all over the World, Leipzig (2014, 2011), 2 Bronze Cubes from The ADC 89th Annual Award, New York (2010), Gold Prize from the GDC China Graphic Design Biennale, Shenzhen (2009), 2 Best Design Awards from The 7th National Book Design Exhibition, Beijing (2009), Silver Prize from HKDA Design Awards, Hong Kong (2009), Silver Cube from The ADC 87th Annual Award, New York (2008), Grand Prize, Gold Prize for Image Recognition and Gold Prize for Publications from the GDC China Graphic Design Biennale, Shenzhen (2007), Gold Prize from The 6th National Book Design Exhibition, Beijing (2004), Multiple Most Beautiful Book Prizes in China, Shanghai (2006 to 2016).

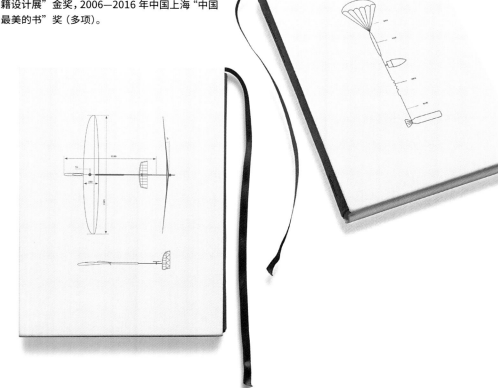

Pupils Walking with Me

喜欢可乐拌冰激凌的乒乓球冠军 —— 我的老师吕敬人先生 | 小马哥

2000年，我大学毕业了，由于我比较听话，按时完成学校的作业，所以我有留京名额。当时有三个工作选择：一个是去某大学教书，一个是去中青社做美术编辑，还有一个是去外研社做美术编辑。我曾咨询现在的设计拍档，也就是当时我的大学同学，由于成绩极为糟糕，大学毕业流窜在北京的橙子先生，到底去哪里呢？其实当时我内心已经决定要到中国青年出版社去。因为有三个人的原因：我知道吕敬人先生和吴勇先生在中国青年出版社，中国青年出版社蓝狮子社标是王序先生设计的。这是一个什么样的地方呐？2000年8月，我终于到中青社报到了。

2000年8月到2001年，我一直坐在吴勇先生工作过的办公室、吕敬人先生使用过的带有透台的巨大办公桌前工作，每天都使用王序先生设计的社标，但我从来没有在社里见过吕敬人先生、吴勇先生，也从来没有见过王序先生，只是不时接到日本邮寄来的给吕敬人先生的资料，交管局给吴勇先生的汽车违章通知……因为他们都把工作挂在社里，独立到自由市场上进行设计工作了。

当时的部门领导寒柏哥看出我的意思，决定介绍我选择几个工作室代职学习。他特意安排我到吕敬人先生工作室送稿件，以接近先生；以工作交流的方式把我带在身后去王序先生工作室参观、请教。他们到现在都不知道其实我曾经做过探路小间谍。看到王序先生很高大、很威严，我一个女孩子有些紧张，加之我了解王序先生广州、北京两地跑，本人不长驻北京工作室，所以没敢请寒柏哥和王序先生沟通学习事宜；而吴勇先生年轻、时髦，很酷的样子，有点距离感；吕敬人先生总是笑眯眯的很和蔼，设计很多传统典籍大书，正是我作为年轻人要补课的，所以就咨询寒柏哥，问吕老师那里他可以帮忙吗？寒柏哥说没问题！但当时的美编室主任因为个人原因，坚决不同意，事情就拖了下来，直到有一天寒柏哥对我大手一挥说：主任退二线了，你现在就去吕老师那里吧！

但第一次给吕老师电话就哭了鼻子，因为吕老师说工作室新来了很多人，没有地方了，我哭了，因为我觉得好不容易来了机会，又不能去了。可不一会儿吕老师电话又过来，预计他听出我哭了吧，哈！所以我从2002年4月带着电脑和必须要独立完成的社里的设计任务，成了敬人设计工作室的一名实习员工。

工作室早上9点上班，晚上9点下班，非常繁忙。当时人员很多，我还记得每一个员工的名字，他们是：德浩、张朋、小梅、雅静、小牛、小柏、小严、刘瑛、小孟、小芮。吕老师在一个小小的办公室，里面陈设简单，但很有书香气息。工作室设计人员使用两个房间，我在稍大的房间靠窗的位置。每天在工作室吃两餐，有一次中饭，我看到一块排骨，样貌很好，想要夹过来吃。正好橙子来了电话，我起身去接，等回来的时候，那块样貌很好的排骨没有了，不但样貌好的没了，那个菜盆也只剩一片孤零零的菜叶。回到家里我抱怨他，从此以后橙子再也不在吃饭的时候打电话了，现在提起我俩总是笑个不停。

一般中饭的时候，大家会聚集起来看凤凰卫视的《锵锵三人行》，记得吕老师喜欢香港的郑秀文和台湾的陶子，喜欢陶子唱的：十个男人七个傻，八个呆，九个坏⋯⋯哇，吕老师是乒乓球高手，午休会和大家打乒乓，没人打得过他，他是工作室的总冠军。我有时候想，是不是别人都是初学，显得他好厉害！先生工作的时候总是小跑、小跑，指导每一个人，时间总是很紧张，先生背影很壮阔结实，浓浓的胡子，无论多忙从来不发火，从来不发火！耐心地一遍一遍讲解。在工作室工作期间，我记得主要负责了两套书的设计，都是关于传统的大部头书籍，当时完全遵循先生的意思，很注意细节设计，他讲解设计思路很开，有的时候盯在电脑前共同设计，有的时候打印出来给他看，每次结束、再开始，面貌都变化了。当时杉浦康平先生和安尚秀先生还曾到工作室来指导。

我印象最为深刻的是先生的设计气质儒雅、样貌浑厚，层次很丰穰。在图片上，多压抽离出来有意味的线，立叶使得多层图片焕发光彩，有很强的设计代入感。在古版书字体没有被电子化复刻之前，我印象里面先生是最早用古版书经典刻本文字做设计的。在工作室里，我们都在先生收藏的绝版古书里面翻找、扫描字体。先生运用古版书刻本字的设计古朴、拙气、书香，很好看。先生设计小书灵动智慧，记得在书店看到一套叫作《生命意识》的书，先生的几笔简笔，加上一个运动的木头小人物，很巧妙。我尤其喜欢先生设计的《经济科学译丛》系列封面，到现在如果让我去设计这样一套书，我仍然不能完成如此有冲击力、理性、冷静又当代的设计，在市面上我也没有看到比这个更好的经济学设计，是我个人认为的先生的巅峰作品之一。

先生提出编辑化设计是对中国书籍设计的重要贡献，倡导书籍整体设计概念，对装帧设计的片面性提出修正，这个方面他多有著述。我是实战主义者，向来不重理论，但却深用这些方法。

在我实习期间，工作室虽然很忙，但也会组织大家去旅游，记得有一次是去密云漂流。当时中午太阳很大，我和先生一条船，他拼命划船，头上晒得红红的，我戴着先生的帽子，也拼命划，旁边船上的雅静喊道：吕老师把帽子戴上！我的脸红了⋯⋯啊呀！吕老师的光脑袋被晒爆了。

2002年10月，我被出版社召回。现在2016年，先生年近70了，依然是那么超喜欢可乐拌冰激凌，那么健康，充满活力，喜欢和年轻人在一起，提携年轻人，赞扬年轻人，写到这里我的眼泪扑扑掉下来！不但对年轻后辈充满善意，先生提到同辈、长辈设计师，无论何种风格都衷心赞誉。

回忆起14年前的往事，仿佛又看到先生熟练的草图勾画，非常洒脱，具有很强的造型力。虽然我的设计风格和先生差异很大，但我到现在仍然使用先生传授给我的他恩师杉浦先生很酷很酷的"微尘"和"杂音"概念，设计了那本由著名策展人欧宁先生编著的《南方以南》。

何明 HE MING

平面设计师、策展人。1974年出生于新疆,现居住生活于四川成都。1993—1995年为自由音乐人,1995年就职于深圳言文设计从事设计工作,2000年回成都创办7788文化传播公司至今。多年来以设计师身份渗透到书籍、杂志的图片、文字及采访编辑工作之中,不断尝试独立出版物的设计创作与发行。参与编辑设计作品有《360观念与设计》《1314》《同上》《原创力》《纸能》《或》《从小到大》《想念亚美尼亚》等纸媒。2005开始介入展览策划工作。从内容策划、空间规划、展览展陈、动线、光源、视觉等多方面通过展览媒介推动设计,艺术,产品与生活、与社会、与人之间的导向传播。参与策划的展览有建设展、随意摄影展、疾风迅雷——杉浦康平半个世纪杂志设计展、高兹·格拉里奇海报展、对话与视觉——李永铨与设计二十年巡展、社会能量——荷兰设计展、7080设计人展、纵目摄影双年展等。

Graphic designer and convention planner. Born in 1974 in Xinjiang and currently lives in Chengdu, Sichuan province. Freelance musician (1993 to 1995). Dealt with design works at Eonmun Design, Shenzhen China in 1995. Returned to Chengdu in 2000 and currently runs *7788 Cultural Company*. Deeply involved in books, magazines and graphics, texts and interview editing and trying to create designs for independent publications and publish them. Participated in editing paper media books, including *Magazine 360 Thought and Design*, *Magazine 1314*, *Magazine Dongsang*, *Magazine Wonchangnyeok*, *Paper Energy*, *Bump*, *From Childhood to Adulthood*, and *Missing Armenia*.
Since the start of convention planning in 2005, he has created exhibition planning, art, product and life, society and dissemination of guidance to humanity in various fields, including content planning, space planning, exhibition display, moving line, light and visual areas. He partook in diverse exhibitions: Construction Exhibition; Photographic Exhibition; Do whatever you like, Gust Trust; Design Exhibition of Sugiura; Gotz Gramlich Poster Exhibition; Dialogue and Viewpoint; Traveling Exhibition of Tommy Lee's 20th Year; Social Energy - Holland Design Exhibition; (Hong Kong, Mainland, Taiwan) 7080 Designer Exhibition; Biennale Exhibition – Look at Photography from a Distance.

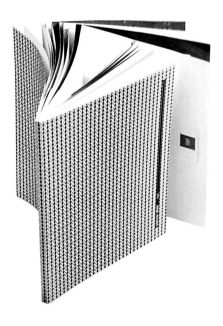

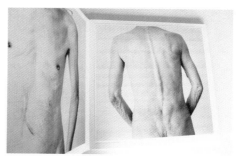

Pupils Walking with Me

目语 —— 敬人师对我的启示录 | 何明

这是光线迷离的某个午后的意绪，鲜妍茂盛繁华的早晨过去了，时间之流渐趋缓慢——去除那些不必要的生活构件。人生旅途似乎也类似一天的征候，痛与绚丽不再是近在咫尺的邻居，它们像飞驰的游牧部落，马镫与刀鞘的撞击声远去；但审美依然敏锐，虽然眼中容不下多出的一块石头。雾霾的季节里适合歌唱，她能带来风和雨。

天大地大，天地间人显得微不足道，于是便有了谦卑，因为谦卑更胜于对信条的坚守。敬人师时时刻刻在塑造谦卑的哲学，不仅在事业上也在生活中。高山仰止，景行行止，敬人师之谓也。

回溯过往，在敬人师的推动下，杉浦康平老师的巡展在中国举办；北京站之后，在敬人师的信任下，我幸运地成为了杉浦康平老师成都站的负责人。与敬人师共事是幸福的，他平实朴素得就像一座山，沉稳，丰富。布展的时候，敬人师事必躬亲，也是那段时光敬人师言传身教，让我受益匪浅，让我从个体的设计师身份转换到设计对于社会的思考，让我意识到展览交流对一座城市、对行业、对社会的意义。也是在那段时光里敬人师在我的心里种下一颗种子，他没有许诺，只是教会我如何培土、浇灌、施肥……敬人师像一个望风而作的农人，却能望穿气候与农时亘古的永恒。

杉浦康平老师的展览之后，我试着放慢设计的脚步开始学习策展，学习如何不断地破坏墙面并重塑墙面，学习找寻榜样作品背后的所思所想，学习交流分享的力量和感动，学习测试光源的微妙变化，学习体验距离和改变观看视角，学习和更多优秀的良师益友合作，学习找寻多维视觉语境的表达方式，我开始对内容对空间锱铢必较。我开始学习剧本、导演、故事、场面、演员、镜头感、站位、画面、台词、配乐、沉默、规范与禁忌……随意摄影展、成都时间摄影展、高兹·格拉里奇海报展、对话与视觉——李永铨与设计二十年巡展、社会能量——荷兰设计展、7080设计人展、想念亚美尼亚——阮义忠摄影展、手不释卷——赵清书籍设计展、纵目摄影双年展、NUAR艺术节……十年间一场展览接着一场展览。

此时设计呢？好像设计离我越来越远……不，其实我从未放下过设计，是敬人师教会了我如何在空间里做书，只不过此时，纸变成了墙、图片变成了地、文字变成了光、图表变成了声音……每策划一场展览就像完成一部书，这部书和城市和社会有关。空间是一本极其精密、庞大、复杂、优美，充满激情、快乐、悲痛的书，一本时间和空间叠加的书，通过展览可以达到这一点。"纵目"是我做的摄影双年展的题目——是对视觉潜意识，对视觉的人类学和伦理学意涵的关注，对城市与当代人生活场景的着迷，注视与失焦，摄影的根蒂。我喜欢在相机的取景框中观察世界，此刻，远古气息与我同在。孟子曰：爱人者，人恒爱之；敬人者，人恒敬之。玉篇：敬，恭也，慎也，肃也。每见敬人师温厚的君子目光夹杂着孩童般的真挚时，我就想到，这不，先生也是纵目呢，纵观古今，打通美学天人之际的深邃之目。

古池 / 青蛙一跃 / 扑通一声。松尾芭蕉的这则俳句为这个简寂的午后做了注脚，也是敬人师在设计之外对我的启示。

韩湛宁 HAN ZHANNING

中国出版协会装帧艺术委员会常务委员,亚洲铜设计顾问公司创意总监,曾任汕头大学长江艺术与设计学院教授、深圳市平面设计协会秘书长。曾获得德国红点奖"Best of the best"大奖,全国书籍设计艺术展览金奖,"中国最美的书"奖,GDC平面设计在中国双年展银奖。

Earned master's degree from the Academy of Arts & Design, Tsinghua University, Executive director of the Book Design Committee of the China's Press Association, Director of Asia Holizon, Professor of Yangtze Arts and Design College, Shantou University, Chief secretary of the Shenzhen Graphic Design Association. Best of the Best in the Red Dot Award, Gold Prize from a National Book Design Exhibition, One of his works chosen as the Most, Beautiful Book in China, Silver Prize from "GDC Graphic Design in China"

我与恩师吕敬人

韩湛宁

1996年9月，亦师亦友的王春声从北京回太原，送给我一本令我醍醐灌顶的书——《书籍设计四人说》，这本书来自在北京刚刚举办的一个特别的展览——"书籍设计四人展"。我并不知道这本书与这场展览在当时只做封面的"装帧"出版时代下所产生的影响，但对我而言不啻是一场惊雷。

敬人入心

《书籍设计四人说》带给我的惊诧与震撼是独立的，与当时它的巨大影响毫无关联。彼时的我刚刚开始热爱平面设计，对当时的书籍装帧了解甚少，但是亦对简单粗糙的甚至连封底都不画的封面设计、书脊与书面永远是波浪起伏的粗糙印制深恶痛绝。我独立的惊诧让我记住这场新风暴的倡导者——"吕敬人"这个名字，以及"书籍设计"这个崭新的概念。

那天晚上，我捧着书，看着这个当时已经享有盛誉的设计家留着胡子的照片，琢磨着"敬人"这两个字。这对于我这个在"建国""援朝""跃进"们的教育下，在"文革""建军""卫兵"们的陪伴下长大的人来说，是多么大的陌生感和吸引力啊。"敬人"这两个字一下子让我联想到"温良恭俭让"，联想到一种从"旧时代"带来的传统的文人气息，乃至谦逊到极致的温文儒雅。除了思考这个温文儒雅的人有着怎样强劲的思想才能提出"书籍设计"这个惊雷般的概念之外，我隐约地觉得，他和他的作品一定有着我向往的那种中国文化脉搏的跳动，以及改变这个跳动的力量。

再见到"吕敬人"这个名字是在一年后的1997年，是在书籍《子夜》（手迹本）上，记不得是在哪里看到的。那时我已经为几家出版社做了一些书籍设计了，故对设计得好的书非常敏感。看到《子夜》（手迹本）时非常激动，那温和而又充满力量的设计令我向往，一直细细翻阅、爱不释手，甚至对书籍的布面函套都摩挲再三，连那蓝色函套上白色的"子夜"两个字，都像子夜的月光照耀一般，让我心底敞亮。这本书也让我记住了出版社的名字——"中国青年出版社"。以至于后来，对所有标有"中国青年出版社"字样的书，我都会多翻阅几次，觉得那样会离吕敬人老师近一些。

等后来我又看到吕老师的《周作人俞平伯往来书札影真》《朱熹榜书千字文》等作品时，看到那毛笔字与宣纸、碑刻与拓印在柔美的书卷上，以崭新的书籍设计方式呈现的时候，我心中所有关于中国人温润雅致而不失刚健雄浑的文化与情感都涌现了出来，甚至是一种欢喜：原来，终于有人将我们中国人的文脉以设计的方式续接了起来。

初识恩师

念念不忘，必有回响。终于在1999年春天，我们——张国田、王春声、赵紫春和我等人——以山西平面设计协会的名义邀请到吕敬人老师莅临太原，出席一个展览并讲学。

初见吕老师我反而紧张，也并没有太多机会可以交流。吕老师真的温文儒雅，还很幽默，一点都没有大师的架子，浓黑的络腮胡子很茂盛，黑色边框眼镜后面是一双永远笑眯眯的眼睛。活动很顺利，参与活动的设计师们都受到了鼓舞，受益良多、感念不已。活动结束后，我们陪吕老师去山西几个大宅院游览时，有了和吕老师较多的相处时间。记得在王家大院，吕老师指着墙上或窗上那些优美的砖雕或木雕的书卷、画轴造型给我看，说："你看，我们祖先是多么敬畏和爱惜书卷啊。"然后微笑着、意味深长地看着我。那一刻，他的神情和我在 1996 年想象的他完全一模一样。

受他的鼓舞，返回太原时，我斗胆拿出一些作品给他看。他对我这个初出茅庐的年轻设计师的作品予以热情赞赏，仔细倾听我的问题，并给予解答和指导。在他返回北京时，诚挚地邀请我到北京去他工作室交流。

过了几个月，1999 年初秋的时候，我当时正在长沙为山西参加全国书市做展位设计和布展工作，突然接到吕老师的第一个电话。当电话里说"我是吕敬人……"的时候，我有些吃惊和激动，不知如何回答。吕老师在电话里告诉我，杉浦康平老师的《造型的诞生》中文版出版首发式在北京举行，同时会做一场演讲，机会难得，希望我可以去听一下。但是由于我在长沙负责布展走不开，吕老师便告诉我说："没关系，我给你留一本杉浦老师的书，你下次来北京到工作室我给你。"这个电话令我欣喜不已。

很快，我就去北京拜访了吕老师，当时吕老师的工作室是在亚运村的欧陆经典。吕老师热情地接待我，并对我带去的设计作品给了悉心的指导。那次我深入地请教了吕老师从"装帧"到"书籍设计"的概念区别，也告诉了吕老师我的困惑："我很难把握在书籍设计中的角色定位，既担心驾驭不了文本，设计表现得太表面，又担心'编辑设计'过度，被指越界。"吕老师说："你有这个担心已经说明你在思考装帧和书籍设计两者的责任担当的差别了，这是一个很自然的阶段。设计师应该是一个导演，将书籍中的文本、图像、色彩、空间等视觉元素进行调度，注入有生命力的表现和有情感的演绎，最终形成既还原于文本思想又有新意的阅读形态。你的困惑，是需要超越设计的知识积累和追求更高的文化素养与文本碰撞，引发触类旁通的新构想，这是必须要面对的过程，而不是装饰美化书衣那么简单。"那次吕老师对我提出了更高的要求，即设计之外的人文素养的提升要求。就这样，恩师予我深情教诲。

渐入师门

之后的 2000 年夏天，我到了深圳工作，相比太原，到深圳后与北京的来往就多了许多。不是我去北京，就是吕老师来深圳，总是小聚一下，并且检查一下我的"作业"，就是看看我新做的设计作品、书籍画册等等，并予以指导，有时间也到我公司给设计师们进行辅导和交流。

记得一次在深圳雅枫酒店，我和好友徐茜去看望吕老师，同时我也带了几件作品去给吕敬人看。因为时间紧张，我们边吃饭边讨论，吕老师一一指点，其中有一本商业画册，做得不够深入，吕老师温和而坚定地看着我说："无论是何题材，既然做了，就用心做好。"我听得汗流浃背，连连认错。事后徐茜笑我说："我第一次看到你这个骄傲的人也有这样狼狈的时候啊！"

这是吕老师善意的力量，即使是批评，也从不形于色。其实这来自他的家教。"敬人"这个名字就与此有关。吕老师出身旧上海的书香之家，自幼家中藏书丰富、气氛欢乐，他的父亲饱读诗书、热爱书画，在事业上励志求精，对孩子们秉承传统的伦理家教；母亲是虔诚的基督教徒，终身行医施善，在淞沪抗战中冲在前线抢救伤员，受到十九路军抗日将领的表彰。他的父亲在 96 岁时，还曾以"敬"字写了一纸条幅："敬人吾儿 铭记：敬事以信、敬业以诚、敬学以新、敬民以亲。"吕老师说："父亲要我牢记做事、做人、做学问的道理，这也是我一生的座右铭。"而当时，书写之后的父亲一再谦逊地向他说道："献丑，献丑！见笑，见笑！"

在很多的日子里，吕老师这些温和、善良、恭敬、俭朴、谦让的品格，都深深地影响着我。记得一次和吕老师在日本，回酒店时已经很晚了，在温暖灯光映照下的巷间林荫道上，两个人边走边聊，我不由自主提高了嗓门，吕老师就提醒我说现在是晚上 11 点多，你声音太大会影响别人的。说得我不好意思，于是压低了声音。回到酒店，大门紧闭，黑灯瞎火。日本的好多酒店很精致而又很俭朴，不像中国酒店般富丽堂皇，且大晚上还灯火通明。到了门口，我直接就上前摁门铃，吕老师发现了，就把我的手摁住说："不要按门铃，你不是有钥匙吗？"因为酒店给我们每个人一套钥匙，有房间门钥匙和酒店大门钥匙。他接着说道："不要打扰别人！这么晚了，服务员可能在休息，你叫人家起来给你开个门多不好啊，天气还这么冷。"一句话说得我脸就红了，赶紧拿钥匙开门。"不要打扰别人"，这句话从此就留在我的心里。

在渐入师门的日子里，吕老师言传身教，将"温良恭俭让"这些我们中国人的品行内核植入到我的内心，让我受益终生。

"求"我读研

2006 年，我开始在汕头大学长江艺术与设计学院任教，教学任务颇重，加之我自己的设计公司涉及诸多领域，工作繁多。吕老师看到我经常忙到深夜，还要开车几百公里去汕头大学上课，非常心疼我的身体；又看到我频频地跨越各类设计工作，不够专一，分散了大量精力，他非常着急。

于是吕老师就告诉我，他年轻的时候也是什么都做，并因此被杉浦康平老师批评。杉浦老师告诉他说，你什么都想做，结果会什么都做不好。好比一个手的五指，都想往外伸，结果是每一个指头都没有力量，而你把五个指头收回来，聚拢成一个拳头，那么你再打出去，就非常有力量了！

吕老师劝我专注一些，同时希望我可以静下心来，于是建议我去国外读书，在设计上深入研究下去。但是，彼时的我又开公司又教书，孩子年纪又小，无法脱身去国外读书。大概两年后，吕老师见我无法出去，就说，你到清华美院考我的研究生吧。我诚惶诚恐，诺诺答应。早在 2002 年，吕老师已经由中国青年出版社调任清华大学美术学院担任教授，并在中央美院等院校担任客座教授，致力于中国书籍设计的教育工作。但是我那时依然忙碌无法复习应考，又拖延了两年，后来吕老师实在生气了，就说，你再不考，我就退休了！这次吓得我赶紧认真复习准备。经过艰辛的努力，终于在 2010 年以优异的成绩考取了清华美院的艺术硕士研究生，成为吕老师的最后一名研究生。此事经常被吕老师调侃说，我是"求"着湛宁读我研究生的。这一直让我感到羞愧。

"求"我读研究生这件事，其实是吕老师对后辈、对设计行业年轻一代提携和激励的许多故事之一。对我而言，通过这件事可以看出吕老师对我的爱惜和期望，期望我更出色更优秀，可以在两年多的时间内心无旁骛，专心研究设计。当时我已经由副教授升任教授，且担任硕士生导师数年。此时再去考取一个硕士学位，已经没有任何功利目的，纯粹是跟随吕老师专注学习书籍设计，这也是我此生最大的荣幸。

陪我采访

在清华大学美术学院学习期间，我除了专心学习之外，还有一个重大的科研项目同时进行，就是对健在的老一辈书籍设计家进行深度访谈和研究。这是吕老师对我的学习和研究进行系统指导的一个重要课题。

其实这个项目早在 2008 年就在吕老师的指导下进行了。每次采访之前，吕老师都和我一起研究史料，替我联系被访者，并且多次亲自陪我前往采访，甚至担任翻译（如张慈中、范一辛等老师均吴音浓重，吕老师也是上海人，就替我当翻译）、摄影等。而采访后又多次审阅指导我的文稿，整理出来文稿后又和我一起再与被访者交流沟通，以取得最翔实的访谈结果。事后，吕老师还指导我对材料进行分析研究，与我一起提出观点，做出扎实系统的研究论著。

记得一次是初夏，吕老师和刘晓翔陪我去上海采访前辈范一辛老师。离开酒店去范老师家的路上，我们发现吕老师衬衣袖子上破了一个小洞，一再提议去买一件新衬衣，但是吕老师拒绝了，他直接把袖子挽了起来，说"这样不就好了吗？你们别再说了，我们直接去吧，不要浪费时间！"我们哑然一笑，就听从他的指示。虽然吕老师对生活有他的要求，但他其实一直是一个俭朴的人，他更多是对品质的追求，而非表面的东西。

而之前采访张慈中老师，吕老师不仅亲自陪同，还邀请宁成春老师一起为我护法，外加刘晓翔兄，可谓阵容鼎盛。张老师也颇为感动，对我们称谢不已。同样，采访前辈曹洁、吴寿松老师，也都是吕老师、宁老师和刘晓翔兄陪我前往数次，这何止是"提携"二字可以说得了的啊。

虽然现在这个工作还在继续，但是 86 岁高龄的吴寿松老师已经在三个月前溘然长逝。悲痛叹息之余，我也庆幸得以在他，以及更多前辈的有生之年可以聆听那些重要史料，也突然意识到，吕老师通过这个方式，让我将过去与传承连接了起来。

最好的时光

2012 年 4 月底的一个下午，那时我还在北京求学。吕老师开车来接我，说带我去一个地方，去看花开。我们走了一个多小时，来到北京昌平的一个山区，看到满山的花在黄昏的阳光下耀眼地开着，美丽极了。这是吕老师常来的地方，房子在山坡上，树特别多，有杏花、桃花、迎春花……各种各样的花，美极了。我想起之前吕老师和我说，要不你春天来，我带你去看花；要不你秋天来，我带你去摘果子。

看着夕阳下花开满目，我陶醉了。山区晚到的春天是最美的季节。吕老师说："花，一年就开这一个星期，花季生命很短，我们应该敬重并欣赏它的美！"当晚我们就在房子里住下来，屋内宁静雅致，充满书香，富有设计感和艺术气息，好奇的我几乎把房间研究了个遍。第二天早晨 7 点起来，吕老师已经把早餐做好了，好生惭愧的我连忙按照吕老师的吩咐和他一起把餐桌抬到外面去。我们在杏花树下吃着早餐，太阳悄悄爬上树梢，满树的花朵在阳光下变换着色彩，我觉得拥有生机的花特别的美，更觉得吕老师是要我珍惜人生每一瞬间的宝贵时光。

记得那天，我们讨论了他的书籍设计作品《中国记忆——五千年文明瑰宝》。获得 2009 年"世界最美的书"殊荣的《中国记忆》是我最喜爱的吕老师作品之一，其设计思想是将中国文化精神最具特点的天、地、水、火、雷、山、风、泽进行视觉图形构成，以体现东方的本真之美。记得他之前曾经和我说过，《中国记忆》是以浏览中国千年文化印象的博览"画廊"作为设计构想的。我就问吕老师，其设计是否与他热爱自然有关，自然的美在他的设计中与文化融为一体，成了这一中国文化的视觉表达？吕老师回答然也。他说，其书封面与目录页等图片就是他在慕田峪长城上拍摄的，图中烟云升腾、山脉与长城融为一体，不正是中国文化"神"与"物"的统一吗？

吕老师说："作为一个设计师也好，作为一个人也好，我们应该懂得敬畏大自然，尊重大自然，我们要有对美的向往和感知的能力。古人不是也讲'师造化'吗？你就安安静静跟我在一起，看花开花落，看日出夕照，看星辰斗移！"

那天特别感慨，吕老师让我真正懂得美，那天的欢喜还有：我终于知道怎样将内心和自然连接起来了。那天和所有与吕老师在一起的日子一样，成为我生命中最美好的记忆。

2016 年 6 月 21 日初稿　6 月 30 日二稿　7 月 1 日三稿

杨林青 YANG LINQING

1999年毕业于清华大学美术学院视觉传达设计系，毕业后工作于北京敬人设计工作室。2002—2006年留学法国巴黎，毕业于巴黎国立高等装饰艺术学院（ENSAD）编辑设计专业。2007年回国后在北京成立个人工作室，从事出版物的策划、编辑与设计工作，并致力于中西文字体的媒介应用和图形信息交流的研究。

Graduated from Department of Visual Communication, Academy of Arts & Design, Tsinghua University in 1999. Worked at Jingren Studio Beijing after the graduation, Studied in France between 2002 and 2006 and graduated from Editorial Design, ENSAD. Founded a studio in Beijing in 2007 and currently focusing on planning, editing and design of publications. Putting in efforts to apply calligraphies of Chinese characters and Western letters to media, and study graphic information exchanges.

一直在路上 | 杨林青

2016年的某一天因工作室搬家，我在整理文件时发现了一封11年前给吕敬人老师写的信的复印件。那时我已在法国两年多，吕老师要我谈谈对书籍设计的看法与体会。虽然这封信记录的是我11年前的想法，但再读时发现自己一直都在这样实践着，越走越接近自己的内心。

记得出国前，我问吕老师能给我一点什么建议，吕老师说："像一块海绵一样，捏紧了，放进水里，再放开。"两年后的一天，我特意给吕老师打了一个电话，在电话里兴奋地告诉他我终于把这块海绵捏紧了，现在终于是放进水里的时候了。是的，在这两年里，除了学习法语，其他所有的时间都在思考关于设计的各种问题。当自己能从以前的那个我走出来的时候，感觉我在国外的学习才刚刚开始。

1999年大学毕业后，在敬人设计工作室的学习和工作是开启我对设计认知的重要起点。在那里，我开始从设计一个封面到设计一本完整的书，开始接触编辑知识，开始重新认识书籍设计，这些都成了我日后继续学习和思考的动力。

下面我将11年前的那封信的内容娓娓道来：

吕老师，你好！

实在不好意思！一直想给你写信，但总是有什么事或因为懒惰才拖到现在，请见谅！希望不算太迟！

上次我回去你让我写一段对书籍设计的看法和体会，其实在回重庆的时候我就写了，但后来没有给你！不是因为怀疑它的好与否而愧于给你，而是当时自己认为很多萌动的想法难以流于纸上，再次回到巴黎时，我开始整理它们！通过这两年在外的游学，我感觉变化最大的就是自己渐渐地走出以往的"平面设计"空间，不只是因为出国看到了更多优秀的设计，而且由于置身于巴黎这个丰富的艺术宝库，其他艺术把我引到一个适当的位置重新思考自己从事的领域，才发现原来以前自己的思维空间如此地"狭窄"。这两年在外也认识了很多朋友，有同专业的学长和设计师，也有建筑、电影和文学领域的。我非常感谢他们的出现，使我把以往与设计并无直接联系的其他兴趣和自己的工作学习紧密地融合起来。我开始站在不同的角度去认识书籍设计艺术。虽然这个认识才刚刚开始，但它却给我一种进一步探索的可能和希望，同时促使我不断地实践去充实它。

书是一个三维的载体，但在我以前的认识里除了这个仅有的立体概念外，脑里几乎全是文字、颜色、图形、版式等这些平面的视觉元素。我会用绝大部分的时间把这些元素放在同一台电脑屏幕上放大、缩小、移动、组合、变化……这是一个有时有趣有时枯燥的"平面"视觉游戏，日复一日的二维生活使自己也慢慢地被禁锢在"平面"的思维里。更有意思的是，我甚至可以把这种平面思维套用在任何一个"平面设计"的领域中：标志、海报、包装、书籍等。实际上这些领域每个都有其独特的载体意义，是不可以相互混淆的。我想在国内时至少我会用"平面设计"一词把它们都一概而论，除了外观功能的差异，其设计的"理念"皆来自同一"视觉"思维。我可以把图形和版式做得"千奇百怪"，但从本质上我认为它们在这些不同的载体中是没有区别的。而这种一概的"平面设计"思维正好可以允许一个平面设计师朝三暮四，避开探索载体存在的意义，而把精力放在可以玩弄的视觉表象上。尽管试图改变，利用不断丰富的材料和迅速发展的印刷技术，但总是在这种表象周围打转。所以这时我并没有真正理解到设计一本书的意义何在，和我为什么要选择做书，要做什么样的书。内心的各种矛盾在两年的激战后渐渐地平静了下来，我开始真正地专注一个东西——书。

所以我想和你谈的对书籍的认识，除了书籍设计本身，还有设计师本人和我们以前提及的书籍编辑问题，以及书籍设计中一个非常重要的要素——时间。我想运用一种立体观的态度把这几者进行逻辑的构筑，虽然这是一个初步的开始，而且涉及种种问题，但我想这种探索将会伴随我一生。

首先，对于书籍本身的立体感观，我经常把书和建筑放在一起比较。因为我认为书籍设计语言从某些程度上与建筑设计的语言极其相通。虽然书籍设计师不像建筑设计师那样相对完全的自主，原因在于建筑创作相对独立。而书籍设计师也将面临这种自主问题，从内容组织到设计由设计师来完成，我想把这个问题放在后面的书籍编辑里谈！首先书籍设计本身与建筑设计到底有什么联系？这种方法虽然显得有些呆板和笨拙，但对书籍空间性的认识来说，我认为却是一个非常踏实的开始。我把所有的视觉要素比作建筑里的砖（像文字）、结构（像版式）、墙体（像纸材）、装饰（像图形）、体态（像外形）。如果我们想象所有实质性的材料都是透明的，就像今天我们看到的很多玻璃建筑，空间分割合理、功能科学而有序、造型独特而优美，由此可以推想一本书里各种元素之间的构成关系和版式结构与书籍形态的相互关联。一本书的立体感官不仅来自这种实质的形态，而且还来自视觉和其他人体感官相互影响、作用后给人的一种整体空间感受。在建筑中，人们可以凭借所有感官感受到物质空间升华成为对人的情感起作用的意境空间，和书籍设计里综合运用一切可以影响人的感官因素以获得内容之外的精神满足是一样的。综合以上因素完成的一本成书，人们对它的欣赏和阅读与一栋建筑作为景观被由内向外观景是同样的道理。在这里我不是非得把书和建筑之间的每个环节都生硬地拼凑在一起，而是这种比较使我能更好地理解书籍这个特殊载体的空间意义，同时也让我很容易脱离出以往的"平面"视角去观察这个人们精神栖息的场所。

那为什么我想谈谈设计师本人这个问题？因为我认为它与后面提到的书籍编辑有着非常密切的关系。在谈这个问题之前我想先讲两段生活中的学习经历和体会，它们是两次内容同样的讲座和四本内容不同的同一本书：①去年夏天在北京和大家一起听了日本设计家福田繁雄的讲座，实际上四年前在工艺美院我已经听过一次，而且和这次的内容一模一样，连幻灯片的顺序都丝毫未变。讲座的前30分钟，我心里很不安。一是已经听过，二是对他的这种重复很不满意。在当时很多人站起来不断朝他和他的作品拍照的同时，我开始平静下来思考另一个问题：为什么他总是一成不变地在讲述这些作品？最后我突然顿悟，对一个年近70岁的前辈来说，他已把余下的精力寄托于处于成长过程中的年轻设计师身上，其实他想对在座讲述的不是这些眼前的"大师"作品，而是在讲述他自己的人生哲理：从一点小的兴趣开始，用尽一生的精力去探索它从而获得一种独特的语言表达方式。兴趣源点虽易找，但坚持难耐！年轻的设计师太需要这种秉性！如果说第一次讲座是一次亲眼目睹大师作品的视觉震撼的话，第二次便是一次与他人生真诚对话的心灵震撼。②在巴黎唯一的视觉设计画廊里，一次偶然的机会我听了波兰海报设计师 Michal Batory 的讲座。他是目前法国非常活跃的设计家之一，经常能在巴黎大街小巷看到他设计的各种戏剧、文化海报。他善于把几个毫无关联的事物放在一起进行重构，而这种个性的视觉表达语言也为他赢得了成功。成功的背后是什么？是什么使他有如此独特的视角？这次讲座为我提供了一个最佳的答案，那就是他喜欢捕捉生活中不被人注意到的"景致"，如一个被丢弃的塑料袋、地上的一摊水迹、一个畸形的土豆等等，然后用自己的图形想象赋予它们新的生命。在整个讲座中，绝大部分时间他把同一本书翻了四遍。这本书收集了他平时生活中捕捉到的图片，在图片上全是他平时的"涂鸦"，这时你会发现，一个土豆变成了一个贪睡的肥汉，一个塑料袋变成了一只忧郁的狐狸，两片剖开的蘑菇变成了一对互相爱慕的情人等。就这样在每本书的同一图片上演示着他这些不同的充满童趣的想象，一本书变成了四本甚至更多。这时对前面的问题就不问自解了！为什么我要向你讲这段经历，因为回到生活的原点，原来这些优秀的设计都滋长于平时生活中的一点小小的兴趣。在国内，我感觉我们年轻设计师与生活只有间接的关系，整天要么对着电脑，要么对着一堆设计书籍，把自己架空于一个真空的生活环境中，缺少真实的生活体悟。不是依赖于从设计中得到设计"灵感"这种近亲繁殖方式，就是把一些传统文化生硬地与设计捆在一起。这不是真实的生活，更不是一个充满智慧的设计。这一点对一个设计师来说最为重要，它直接意味着设计师自己独特的视角和个性的表达。

那设计师本人与书籍编辑是什么关系呢？首先我认为将来的书籍设计师应该等于编辑加设计。对于设计师来说，"编辑"有着两层含义，第一是书籍内容信息的编辑设计，它属于书籍内在各种体例的逻辑构筑，第二则是书籍原内容的编辑。在国外，我看到很多设计师把原内容的编辑纳入自己的工作范围，而且这些内容已逾越了我们自己的专业领域，我想这种尝试离我们并不遥远。不能说所有的书都必须由设计师自编自设计，但这种编辑将必然成为工作的一部分，这种自主行为必将使设计师更主动地进行创作。而无论什么含义的"编辑"都应具有"思想"，一种编辑思想是否有趣，视角是否新颖，表达是否独特皆来自设计师自己真实生活中不同的兴趣原点，以此去创造形式丰富、内容各异的书籍。

最后我想简单说一下我对书籍另一要素——"时间"的认识。我认为这是一个相对独立的设计要素，一个三维的物体需要"时间"的介入才能被称为一个立体空间，书籍当然也不例外！并且我想作为设计要素的时间应有被动和主动两种概念。对于这点的理解，我又不得不把建筑拿出来作一个简单的分析。在建筑设计中，被动的时间要素源于结构的稳固性和材料的持久性，而主动的时间要素源于设计师把各种自然元素如风、水、光、天气、四季变化等引入建筑中，使其不仅与环境和谐，而且可能以此调节使用功能和人的情绪变化。虽然书籍不能等同于建筑，但同样具有这两种时间要素，而后者——主动的时间要素似乎在书籍编辑中显得尤其重要！有时我在想：为什么在其他媒体发达的今天，书籍这种载体的生命力还如此的强盛？我认为"时间"起了很大的作用，因为在利用其他媒体的时候，人们往往是被动的，也就是你的时间受这些媒体的控制，而书籍正好相反，人们可以自己控制时间。这种自控性为书籍设计提供了一种利用主动时间要素进行设计的可能。在书籍信息的编辑设计中，我们是否还应该考虑到这个因素？目前我也正在寻找可能性，这里只是提出来和您讨论讨论！

以上是我对书籍设计的几点认识，我想还需要大量时间去不断充实它们，所以不足之处请吕老师指正！本来接下来是关于我申请学校的研究计划和曾对您说过的编书计划，看来这个信封不同意！所以只好作罢，内容有点多，对于这两个计划我想迟些时候再寄给您，请见谅！

愿您、德浩老师及工作室其他同仁工作愉快：）身体健康：）

学生：杨林青
2005年1月5日写于巴黎

马仕睿 MA SHIRUI

平面设计师。1979年出生于北京,2003年毕业于清华大学美术学院书籍装帧专业,2005年成立 typo_d 工作室,现工作生活于北京。typo_d 工作室主要致力于公众出版领域的设计,与众多出版机构合作,试图将更开放的设计观念与形式介绍给大众消费市场。
《我兔斯基你》获得第7届全国书籍设计艺术展览最佳作品奖,《香港三联出版社青年作家比赛系列》获得第8届全国书籍设计艺术展览最佳作品奖,《京都历史事件簿》获得2014年"中国最美的书"奖。

Graphic designer, born in Beijing in 1979 and majored in book design, the Academy of Arts & Design, Tsinghua University graduating in 2003. Established studio "typo_d" which focuses on public publishing and plays the role of backbone in distributing a broader design idea and format to the public and the consumer market, in cooperation with publishers. Grand Prize from the 7th National Book Design Exhibition for his work "I Tuzki U", Grand Prize from the 8th National Book Design Exhibition for his work "Young Writer Competition Series of Hong Kong Joint Publishing", His work "Document about Kyoto" was selected as the Most Beautiful Book in China in 2014.

Pupils Walking with Me

敬人与书籍设计与我 | 马仕睿

1999年,我初入大学校门,念的是装潢系下的书籍装帧专业。报考这个专业一半的理由是它相对冷门一些,但仍从属在当时的热门学系之内;另一半是因为我那时还不知道平面设计的概念,一心只想画画,觉得给书画插图也是不错的营生。

那年工艺美院刚并进清华,在第一学年的"基础课"上,大量的时间还是在美术的范畴下进行绘画等造型训练。而后的"专业课"又以当时炙手可热的所谓"创意"作为诉求,琳琅满目的课程蜻蜓点水般地流过。我记得那时的学风是特别推崇标新立异的点子和卓尔不群的思维方式,于我而言,书籍装帧设计这个专业并没有什么存在感。可以想象,如果就此下去,我还会在相当长的时间内,认为自己应该去追求绘画艺术;也可能最后我还是会从事平面设计,但我依旧会以为做设计需要的就是造型和审美,以及灵光乍现式的创意……这对我,必定是歧途。

大三那年,吕敬人老师开始在清华美院任教。两周的书籍设计课程依旧是匆匆而过,但吕老师却把三根神奇的锥子扎进了我心里。

第一锥

吕老师一上来就向我们展示了一个无比奇妙的书籍设计"现场",那些关于杉浦康平、菊地信义等的设计案例以及观念让我对于"书籍设计"有了一丝明确的感受。

吕老师告诉我们,做一本书的设计师,并不是做一个粉饰封面的画师,也不在于要憋出一个石破天惊的"创意";一本书的设计师应该是一个导演,通过既定的素材,去建构一段精彩的叙述——从此我在心中把逻辑、编辑和设计紧紧联系在一起。这对于当时的我是一个炸雷,是一个极其清晰、精准和生动的描述,更是一种救赎,我很庆幸自己在大学三年级的时候就借由这个观念开始建立对于平面设计的世界观。

第二锥

今天的吕敬人先生以书籍设计家的身份而闻名,但20年前的吕老师还曾是一位厉害的插图画家。直到去年我才在吕老师的工作室看到先生早年的书封作品中的一本《无头骑士》,顿时兴奋不已。这本书我家当年是有的,我儿时就时常端详这个封面!一种类似葛饰北斋《神奈川冲·浪里》式的构图方式,着条纹衫的男子背影占据了画面的主体,随着他的面向,视线落到那位神秘的无头骑士身上,蓝灰色的调子预示着一个悬疑诡谲的故事。

这幅画面在我内心顽固地存在着,对我的影响很深,我常能毫无缘由地在眼前浮现出这个图景:笔刷质感、色彩肌理,甚至是书名的竖长字体,以及月下奔骑的无头骑士。我本以为这就是我众多童年影像之一,再无探查出处或再见的可能与必要。谁知,这竟是吕老师的画作!而后,我又有幸观览了先生的很多封扉和内页的插图,都是颇为精美无比的杰作!我才知道吕老师在绘画方面的水平和成就竟然如此大。

◆《无头骑士》　　◆《神奈川冲·浪里》

然而,能做出如此杰作的吕老师却放弃了插图绘画,一心只做一件事——书籍设计。

当年的课上他给我们讲述了杉浦先生对他说过的一番道理:张开手有五根手指,样样皆能;但攥紧拳头专攻一处,才能击出更大的力量。

我那时代仍有不少人有着类似背景:本是追求绘画或雕塑等纯美术,但出于务实的考虑或因水平略逊,不足以入选美术学院的艺术科系，只得转而学习设计。我们总以为设计不过是实用美术，纯艺术学科才是更高远的追求。于是我的人生规划：以设计作为工作来养家糊口，再在闲暇中坚持自我创作，追寻艺术理想多难，一个人即便终其一生的时间精力也未必能深入一个学科的内核，更何况分心二用乃至多用，最后不过是碌碌无为、草草收场而已。吕老师从日本回国后便专心设计，舍去旁枝，把所有的精力都投入到一件事当中，我摊开手掌又握紧，觉得之前自己妄图面面兼顾的想法实在是荒唐。

这两根锥子，救我于求学路上。第一锥使我对设计的观念大转变，发觉了逻辑与编辑这两个概念的神奇与乐趣；第二锥从此将我与平面设计钉死在一起。

第三锥

毕业后两年，我开始以独立工作室的形态进入出版领域。我的工作室的一大特色就是拒绝单独设计封面，而只接受整本书设计的委托，即便是一本纯文字书。而传统观念中，一本书，特别是一本文字书，只有封面这个门面需要由设计师或美编来单独设计，内文只要按照一个既定样张由廉价的排版公司大规模生产就可以了，所以文字书的内文经常被忽视，客户所有的精力都在和封面较劲，一开始很多客户都无法接受我的这个要求。但我们一直不放弃，终于逐渐获得一些机会，而全部由我编排的每一页内文更加契合封面设计中的起承转合，也获得了好评，甚至这个模式逐渐被更多客户们认可，现在虽然还不是普遍现象，但将文字书交由设计师来整体设计已经不是什么稀罕事了。

没错！书籍的整体设计思想就是当年吕老师扎在我心上的第三根锥子，成为我一个根深蒂固、理所当然的观念。

早在1996年的《书籍设计四人说》上，刚回国的吕老师就明确提出了设计一本书并非只是为书画个封面，做个书衣包装而已；而是应该自内而外，整体性地对一本书的各个部分进行设计。同时，引入杉浦先生对于书的五感的描述，将书籍设计立体化。为了明确这个区隔，吕老师开始强调使用"书籍设计"这个词，而回避装帧设计这个说法。

对于整体设计这个观念，我在课堂上刚听到的那一刻起就深以为然，作为一个设计师当然要对被设计的对象的所有细节进行考虑和设定。终于这个观念在我的实践中给予我莫大的助力，甚至成为我的标签。

时间飞快，转眼我做书也已经12年了，对于千篇一律的项目也开始感到疲惫——我开始对"设计"到底是为了什么产生许多的疑惑。

吕老师在2013年从清华美院退休后开办了"敬人书籍设计研究班"。研究班从世界各地请来风格各异的设计师，都是业内传说级的人物，他们带着丰富多彩的思想与经验，每每让人大开眼界。我参加了研究班第三期的学习，并且从第五期开始吕老师让我为研究班的学员分享自己的设计经验，我开始为准备讲座的内容而梳理自己的设计轨迹。在不断地审视自己的设计并追问自己的思路的同时，我竟然发现了自己的设计观念！原来我自己设计的源头和方法就是编辑＋设计，我所关注的一切题目：观念、图形、字体等最终都是为编辑逻辑服务，这是我的潜意识里的想法。"书籍设计"不是我的标签，它就是我的设计方法与思考路径。

就像吕老师多年前讲的，一本书的设计师像一个导演，通过既定的素材去建构一段精彩的叙述。这就是我的设计方式，任何一个门类的设计都是如此，分析逻辑，整理编辑，最后形成一段有效的叙述。同样的一段话贯穿了10年的时间，又一次给予我启示，先后给了我两次焕然一新的救赎。

记得"第8届全国书籍设计展"后的一次座谈，吕老师感叹时间太少了，还有很多想做的事情，但是年岁不饶人，勉励大家要珍惜时光、珍惜生命。

吕老师40多岁才在日本接触到当代书籍设计的观念，醍醐灌顶，从而放弃出版社的工作，成立中国第一间书籍设计工作室，改变了出版行业对于书籍设计的认知。现今从事书籍设计的独立设计师与工作室多如雨后春笋，已很难想象先生当年创立工作室所需要的魄力是何等巨大，那是对一个坚固的认知体系的挑战。

吕老师成名三十载，早就被世人誉为大师，但先生的设计形态和面貌的变化却从未停歇。吕老师的设计体系一直在不断地深入和革新，不曾固化——毫无对自己成功风格的包袱，只看到先生对设计的热忱与探索和更胜年轻人的创造力和革新力。

去年，吕老师带着我们在韩国办展，诸多筹备工作林林总总，大家常会在先生的工作室碰面。每当我们和老师核对进度安排时，吕老师那满满的日程表总是令我惊叹！我真的好羡慕，老是在心里默默祈祷，希望我也可以像吕老师一样，永远不停下工作的脚步，越设计越犀利！

郭琮 GUO CONG

2005年毕业于清华大学美术学院视觉传达设计系，翻译作品有《信息设计》《迷雾与冲突：探究信息设计的跨学科性与方法论立场》。现为法国Aquafadas数字出版中国区认证培训师、自由信息设计师、策划顾问。
曾获得全国书籍装帧艺术展览探索奖银奖，中国学院奖最佳创意奖，2010年英国伦敦中央圣马丁艺术与设计学院信息设计硕士学位与年度最佳毕业生奖。

Graduated from Department of Visual Communication, Academy of Arts & Design, Tsinghua University in 2005. Translated books: Information Design, Fog and Collision; Interdisciplinary and Methodological Stance. Currently working as an instructor in charge of China at Aquafadas; freelance information designer; and planning consultant.
Silver Prize from the National Book Design Exhibition while in undergraduate course, Creative Prize at the "China Academy Award" and other prizes. Graduated from Central Saint Martins College of Art and Design (Magna cum laude).

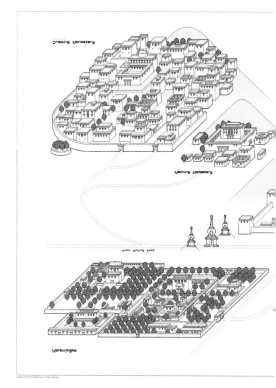

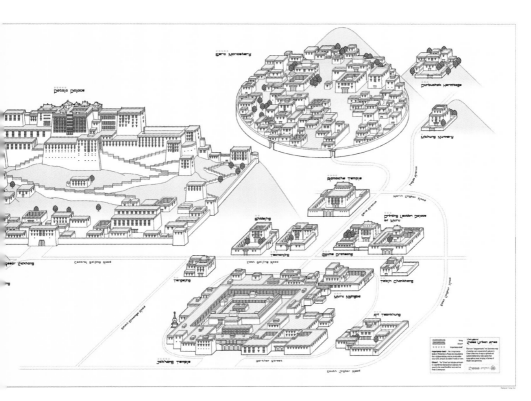

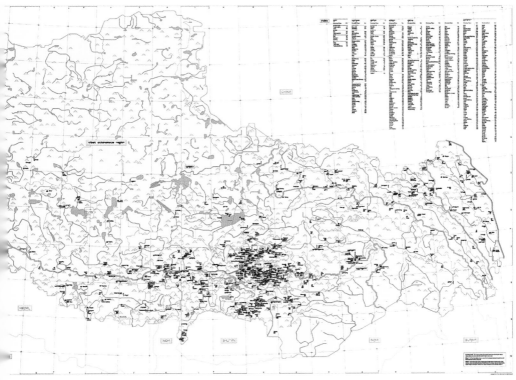

Pupils Walking with Me

执回忆笔 画先生像 | 郭璇

打开笔记本电脑要写吕敬人老师,却望着屏幕迟迟敲不出字来,并不是无话可说,反倒是开关一开,与先生的过往影像便止不住放映了起来。对于患有选择困难症且怵于下笔的我来说,哪些故事更能表达出自己想说的话就成了会令我满头大汗的难事了——想记先生的培育,言传算培,身教是育;想说先生的恩情,学识上有恩,人生中是情……孰深孰浅,哪些说得明白,我道行不够,拎不太清,总担心再多的好故事,放在我这个山寨导演手里,都得拍成烂片儿。罢了,扣上屏幕,擦擦汗,文章依旧空白。

在笔记本电脑如此这般开合了数日之后,自觉不能再拖,想着自己小设计师一枚,何苦跟格子们较劲呢,于是有了念头:给老师画幅像吧,用自己尚且学过的信息图表,给先生画一幅属于自己回忆的图吧。于是,随着画笔也挤出了以下零碎的文字。

认识老师之前,先生大名是久闻的,只是对于一群大二的学生而言,自带"大师"光环的先生,我们除了敬仰,也没有更直观的感受。直到先生教我们第一堂课,我才算记住了,吕老师发光的头顶下,是他的慈眉善目。如果我说这一课改变了自己的人生,可能显得有些刻意而矫情了,可我必须在画这部分的时候陈述清楚,先生的眉目"传情",为我展现了新的天地,也让我之后踏上了出国求学的路。这堂课是当时极具前瞻性的信息图表课,先生更是用了少见的授课方式:一上课就在教室里放映电影《罗拉快跑》,看毕先生为同学们解释了逻辑线索的重要性,并让大家从《罗拉快跑》的影片中梳理出时间与事件的逻辑关系并设计成信息图表。这对于我们这些天天以创意艺术家自居的小愤青们来说无疑是陌生的——逻辑?不应该是理科生学的吗?于是,课程一开始,大家的茫然与迟迟无法上手的作品便成了问题,而先生永远微笑而眯成缝的双眼则成了这个时间点定格的画面——这双带着温度的眼睛不停地在教室中同学们的面前飞来飞去,一位位耐心地引导、鼓励。尽管隔着镜片、眼睛能透光的缝隙也很窄(约为1毫米),但其无限热情深深打动大家。成果自不必多说,大家收获颇丰,而我于此开始领会逻辑思维对于设计的重要价值以及信息图表的特殊魅力。此外,大家都记住了那双永远充满肯定和慈爱的眼睛。

再受教于先生是在大三的书籍设计课程上,先生鼓励大家天马行空地去表达自己所理解的"书籍"形态,并事无巨细地参与到每一位同学的创作过程之中。一如既往,大家的战斗热情跟随着先生的积极互动,一同燃了起来,各种奇特的想法像云朵一般在教室里咕咚咕咚地冒了出来——有能容下一个人的小房子书,有全透明的童话故事书,还有纯手绘的卷轴书……而愚钝的我,看着大家趣味十足的表现方式,却迟迟没有找到自己满意的"书籍"形态,只好硬着头皮把整理好的内容素材打印出来,陈在桌上,等待先生的"讨伐"。"有意思哎,有意思……"伴随先生糯糯的声音,我看到了熟悉的眯成缝的笑眼,本以为会被先生嫌弃,没想到先生嘴里吐露出的和他的眼神一样,是从不吝啬的赞美和鼓励。"吕老师,我想,把自己平时写的小诗歌、小短文,还有平日里拍的照片、画的小画,整理成一本给自己看的很小的书……"我唯唯诺诺解释着不成形的内容,"但是我还没想好怎么做,这体量是不是太小了?"老师推了推眼片,半俯着身子眯着眼看得更仔细了:"挺好的挺好的,内容很有意思……你在考虑书作为这些内容的载体时,也让它这么像你就更有意思啦。"说罢,先生干脆细细讲起书籍五感说了——书与人的互动可以通过各种感官来实现,不仅是视觉,还有触觉、听觉、嗅觉、味觉……一下,前面我都理解了,嗅觉和味觉是啥情况?这一问,先生直接小步跑去从自己带的小布包中抽出一本厚书,卷成一卷就跑了回来,在我面前轻轻翻动:"呐,你闻闻,书的嗅觉。这些纸张都是来自植物的根茎叶,带着自然的气息、油墨的味道,还有文化、历史的味道……"说罢自己把鼻子凑近了书卷闭着眼使劲嗅了嗅:"嗯,这本书比较老,气息不太一样的,哈哈。""书的味觉,是要品的,是从它的内容中去体会它所传递的独特的意蕴。你的小书虽然还没成形,但是内容很有你的特点,好好做,会出来一本很有'味儿'的书!"先生又咧嘴笑了。听先生说话是绵软而微甜的——就像棉花糖。从这些与先生的交流中,我理清了这本书的脉络,也找到了以五感为灵感的书籍形态——就以"我"为主题,用带有天然气味的松木块加上传统经折装,在木块的各个面把关于自己的图文内容编排了进去,书合上的时候组成了大写的英文字母"I",打开后又像积木一样可以堆砌把玩,配合里面稚嫩的文字、儿时的照片，让阅读的人可以体会到"Little I——小小的我"这个书名。这本小书拿到了第6届全国书籍装帧艺术展览的探索奖银奖，而关于这本小书的回忆却让我画出了先生轻嗅书香的鼻头和柔声细语的嘴。

毕业后的三年间，我一直与先生保持联系，先生常会嘘寒问暖，我也常拉着在北京的小伙伴去工作室做客。每每此时，先生就会一路小跑着给我们准备些茶点和饮料，而先生自己的最爱是可乐，捋着胡须喝着冰可乐也算是先生一个有意思的形象吧。当然，每次的重头戏，是围观先生一摞摞抱上书案的新书，先生会激动地挨个跟我们介绍每本书的精彩之处。一次，当我与先生讲述了前往英国中央圣马丁艺术与设计学院研读信息设计的想法后，先生是特别开心的，记得那天也是先生一路小跑着拿来两罐冰可乐，"啪"地打开后，先生就聊开了对于信息图表的理解和杉浦康平先生的图表大作，末了便爽快地给我拟写了推荐信，还一再叮嘱留意好的信息图表书籍。先生认为国内这一块正逢启蒙期，缺乏可以学习的资料，这也成了我在留学期间挤时间翻译《信息设计》一书的动力。这一留学又是近三年的光景，其间一次回来，去大学里拜访先生，先生拉着我在系里看学弟妹们展示在橱窗里的新作，这其中印象最深刻的是先生带领的几位研究生自己手工制造的艺术纸张——手工纸无可替代的质感，朴实而又珍贵的肌理，都在静静诉说着先生及伙伴们对于纸的依恋和对造纸工艺的推崇。以上这些记忆的碎片，用来画就先生的络腮胡子再合适不过了。

待到读完研究生，我一回北京便抱着研究生作品《Lhasa Project —— 西藏地图集》去了先生的工作室，先生眯着眼细细看了整套作品，细细问了设计过程，细细探讨了英国作为信息设计起源地的历史和对西方高等教育模式的想法，末了就笑眯眯地邀请我回学校和学弟妹们分享这些所得，我自知经验不够，便建议邀请当时在英国中央圣马丁学院负责信息设计课程的玛丽亚老师。当得知玛丽亚愿意来校讲学后，先生无比激动，他认为国内的设计领域很需要信息设计的系统理论知识，"得多听听那边的声音"，先生说时在耳朵旁横着划了几下。之后就拉着我去学校商量具体事宜，希望扩充为一个月的完整课程而不只是一两个讲座。这自然不太符合学校规定，外籍教师的课时安排和生活起居对于院系来说都不好处理。"没关系，让玛丽亚老师用我的课时来上课，有三周，我们自己去找宿舍住……"先生一改温柔的形象，回话干脆，表情坚定。待到玛丽亚老师来时，我才知道为了这次能让学生们听到英国原汁原味的信息设计课程，先生牺牲了大量的自己的时间和精力，从课程的设计、教室和课时的调配，到食宿的安顿、玛丽亚的整个北京之旅，甚至是教室里的每张桌椅、投影仪的摆放，全是先生亲力亲为，除了细致，更是强烈的责任心。课前，先生跑前跑后；课上，先生却比任何一个学生都认真，静静坐在台下托着下巴，侧着耳朵听着，不时做着笔记。先生与玛丽亚老师一同跟进了整个课程，作为助教的我则承担了主要的交流工作，大家都受益匪浅，不仅本科生完成了三周的课程，还有许多研究生和其他院系的学生都加入其中。课闲时，玛丽亚受邀拜访了先生的工作室，当看到满屋的书籍作品时，激动得目瞪口呆，还执意买下了先生的整套《怀袖雅物》，说抱着坐飞机也得带回伦敦（严重超重）。此时正值元旦佳节，探究课业之余，先生还亲自下厨为玛丽亚和我们做了道家乡菜红烧大虾助兴。玛丽亚在走时拉着我说，认识吕老师她才理解了东方人的谦逊，吕老师成就如此之高，还如此认真地来听她的每一堂课，如此诚恳地交流自己的想法，令她颇多感触。这些画面，正好构成了先生那愿意聆听、不断接受着新讯息的耳，耳根虽软，对于学术却非常地执著，也正如先生的大虾——看似软乎，却带着韧劲。

之后，先生办了自己的研究班，不计成本地把世界各地优秀的设计老师邀请到北京进行交流与教学。先生也毫无保留将多年积累下来的设计心法——传授给全国各地的学员们。而我，有幸辅助先生在每期研究班讲授信息设计课程。其余的时间我都在陌生的领域闯荡自己的创业项目，尝尽个中酸甜苦辣，在理想与现实中彷徨着，一直看不清前路。一晃到2014年马年春节，正逢内心最为煎熬之际，先生发来了短信：愿今年的马儿，慢点跑，慢点跑……行实事，别奔跑，踏稳步……简短的几句话，是迷雾中的灯塔，感动之余，让我也缓过神来，重新开始审视自己的人生规划。之后又一个傍晚，有机缘于先生书屋喝茶谈心，问起此事，先生一如以前地微笑，缓缓说道："这么多年来，我自认为只做了一件事情，就是做书，我庆幸自己只做了这件事，也用心去做了，所以得了些认可，得了些佳誉，重要的是，生命因此充实了幸福。你现在看得多，可做的也多，但得想清楚有没有哪件事是让你可以一辈子去坚持的……"记忆中的这个场景在我脑海中一直清晰可辨——桌上陶制茶杯旁是先生刚做完的新书《我属猪》，落日撒着余晖，映着先生脑后的头发泛着金黄的光泽，透出的是先生在岁月中沉淀下来的大智慧，书后记着先生的一段话"本人属相：猪也。愚，命薄，且经常成为人们舌尖上的谈资，予人类有点贡献，也算没白活。"敦厚勤实，知足常乐，就用先生教导的人生哲学画了先生的华发吧。

画着写着，先生之像就这样完成了，虽显琐碎稚嫩，却都满载回忆与真情，希望看到的人，能随着笔尖看到一个栩栩如生的先生。

最后，学句先生家乡话：谢谢侬！

PS：阿拉以后少喝可乐好伐啦？

李让 LI RANG

艺术创作者,2007年毕业于清华大学美术学院视觉传达设计系。自2005年以来从事书籍设计及插图工作,设计书籍百余本。
曾获得2009年度"中国最美的书"奖,第7、第8届全国书籍设计艺术展览之优秀书籍设计奖、插图佳作奖、插图最佳奖(七项)。

Artist and creator. Graduated from Department of Visual Communication, Academy of Arts & Design, Tsinghua University graduating in 2007. Has been working mostly on book design and illustration since 2005 (designed over 100 books).
Most Beautiful Book in China in 2009, Seven awards, including Good Book Design, Good Illustration and Best Illustration, at the 7th and 8th National Book Design Exhibition.

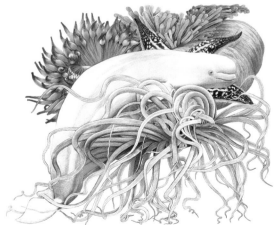

此人爱吃炸猪排 | 李让

1

"老师,您有时间吗?想请教关于书籍设计的问题……"我抱着电脑杵在那。他刚咬了一口三明治,眨了眨小巧的丹凤眼,挥手示意我坐下。

那是大学第二年,编辑在网上看到我的画,问可否为她的书配插图,我表示出信心满满的样子,请她把书籍设计的工作也交给我。实际上,我还没有上过任何书籍设计课程,连排版软件都不会用。当时只有一个想法——我想做书。

然后……几分钟前,我出现在了美院二层的咖啡厅。

在与编辑和自己交战十几个回合后,面对这些不满意的封面我不知所措,心神不宁,抓耳挠腮。想去询问吕老师的意见,但他并不认识我。如何是好?在公开讲座结束后钻进教室问?或者在路上拦截?还是……就在我胡思乱想进行脑内活动的时候,那人闪现在咖啡厅,正在点单!

"老师,您有时间吗?想请教关于书籍设计的问题……"我已经站在他对面开口说话了,甚至还没来得及鼓起勇气。"是什么问题呢?"他马上放下只吃了一口的三明治,同时在极力思索我是哪个学生。我简短介绍了自己(实际上我紧张得不记得当时是不是介绍了自己……),打开电脑给他看那些不同平面组合的封面、插图、文字摆放,诉说自己那芝麻粒大小的烦恼。当时很怕会收到"做的什么玩意儿"的眼神或是批评的话语。他看到那些画,露出和蔼温暖的微笑,说:"这些图都很好,是具象的图形,可不可以尝试抽象的图形呢?也许虚实交错会更有想象的空间吧……"

这句话就像传说中老和尚在小和尚脑袋上敲了一棒,令我恍然大悟。比起这些让人开窍的指导,影响我更深的却是那仿佛可以融化一切的微笑。之后,我设计了人生中第一本书——《芒果街上的小屋》,至今已销售十年,销量过百万,插图深受读者喜爱。而他便是领我走上做书之路的恩师——吕敬人老师。

2

老师知道我喜欢满世界溜达,每次来电话都先问:"李让,在哪个国家呢?……"然后便是嘱咐我记得带新做的书给他看。那是我做书的第三年，这次我拿来了《失物之书》。

他捧着书左翻右翻，脸上挂着微笑，那本书一定也很开心吧，有人面带微笑地在看它。《失物之书》的封面展开是一张大画，原画是整开尺寸，用黑色 0.1 毫米的针管笔一笔一笔绘制，足足花了两个月。而图画在一开始就设计好被分割成 12 小张，可以独立成画，作为书中的插图。现在回想当时制作的过程还有点眼晕的感觉，而比起晕，与出版社扯皮更让人心力交瘁。文化事业是一场又一场的战争，没有输赢，只有不断地冲锋陷阵，而很多时候你并不知道敌人是谁……

"老师，大众书不好做啊，太难了……"我低下头说。他忽然抬头，把眼镜推高，凝神一皱小眉，说道："但是要坚持啊！如果有两条路，选难的那条走。也许你向前走三步，困难把你推后两步，那也是向前迈进了一步啊！如果所有人都觉得大众图书不好做，不做了，那市面上就没有好书了，大家看不到好书，这个行业又怎么会好呢？一个人的力量很小，如果大家都在努力，这力量就很大……"他说完，眯眼一笑，轻轻拍拍我的书说："真好！"

怎么会有这么积极的人呢？还说这么鼓励人的话，于是我又回去继续战斗了，每当要倒下的时候，想起那张笑脸，又不好意思倒地了（笑）。

3

"李让，在哪个国家呢？……"老师又来电话了，喊我参加第 8 届全国书籍设计艺术展览。这位先生思维活跃，精力充沛，忙个不停，远远超过当下很多年轻人（比如贪玩又爱睡大觉的我）。除了做书，老师还经常组织展览活动和教学班，对年轻人从来不批评，总是耐心引导、启发、鼓励。做书不易，我想，能坚持做书的人，大多是因为热爱。小爱暖自我，大爱照四方。

若说是"教诲"，老师给予大家的更多是"影响"。那个人在满腔热情一脸笑容地做书呢，你便也撸起袖子跃跃欲试；那个人在真诚地欣赏他人，看到他人的闪光处呢，你便也积极起来，多多赞赏别人的好；那个人不厌其烦地为别人排忧解难，你便再也不能冷漠，自发去做助人之事；那个人正夹起炸猪排，吃得满脸笑容，你便也伸筷子过去……

有一天我收到陌生人的来信，说她高中时非常喜欢我设计的《芒果街上的小屋》，还有那些插图，觉得做书是很美好的事。现在她大学毕业了，想从事书籍设计的工作，来询问我的建议。我脑中忽然闪现的是老师和蔼的微笑，当年一个小小的微笑，成了深深的鼓励。我给她回了信，信尾放上了一个微笑的表情。

平时像大猫一样独来独往的设计者们，聚在一起有说有笑，聊着书和做书的趣闻。师兄说去找了大师算命，"大师说老师是活佛转世……"他一脸认真和崇拜。我问："那大师说你是什么？"他表情复杂地说："大师没说……可能，我上辈子是草根什么的吧……"大家咯咯咯地乐出了声。

2016 年 2 月

李旻 LI MIN

清华大学美术学院硕士，现任中国日报社资深美术编辑。为报纸、杂志、书籍绘制了大量优质插图。曾获得 2008 年第 7 届全国书籍设计艺术展览插图优秀奖，2010 年第 21 届中国新闻奖新闻漫画银奖，2010 年作品《重负》入选第 1 届"漫画民生"作品展，2010 年作品《共赢》入选 WTO 全球漫画大赛，2012 年首届全国科普漫画大赛二等奖，2013 年第 23 届中国新闻奖国际传播三等奖，2013 年第 8 届全国书籍设计艺术展览插图类佳作奖，2015 年第 23 届中国新闻奖新闻漫画三等奖。

Earned master's degree from the Academy of Arts & Design, Tsinghua University. Edited fine arts works at the China Daily. Contributes quality illustrations to newspapers, magazines and books. Good Illustration Prize from the 7th National Book Design Exhibition in 2008. Silver Prize from the 21st China Newspaper Award in 2010. His work Heavy Burden was carried in the works collection of the 1st People's Life in Comics in 2010. His work Win Win was nominated in the WTO Global Comics Competition in 2010. 2nd prize from the 1st National Science Comics Contest in 2012. 3rd prize from the International Award of the 23rd China Newspaper Award in 2013. Best Illustration Prize from the 8th National Book Design Exhibition in 2013. 3rd prize in Newspaper Comics from the 23rd China Newspaper Award in 2015.

IMAGE

SOJOURN

MY DAYS IN ENGLAND

By LI MIN » limin@chinadaily.com.cn

From February to June, 2012, I worked at the China Daily's London bureau in the United Kingdom.

During the four months, I found London to be a city one never gets bored with. It has a rich cultural heritage and fantastic attractions. It is also a city of contradictions, and it changes depending on the angle from which you view it. I tried my best to record what caught my eye with my pen, and those beautiful moments will always be in my mind.

There are many bridges over the Thames, the most famous of which is surely Tower Bridge. I enjoyed looking out at it. It's like an old man quietly watching London as it changes. London is justifiably famous for its parks, some of which are even a little wild, with deer grazing and foxes stalking the nearby streets at night. Bold squirrels expect food from passers-by. Some of the parks seem like a garden leading to the castle of Sleeping Beauty. But the gardens rarely sleep for long as people will relax on the grass or play on a sunny summer's day. Some of my most memorable moments were spent in the English countryside. When I close my eyes, I can still see the rolling hills with their green grass and scattered groups of sheep and cattle, and I can still feel the spring breeze and smell the sweet fresh air.

我的老师　｜李旻

手机信息提示音响起，我放下画笔，看到的竟是吕敬人老师的一条短信，忙回了电话，挂了手机的瞬间想起已有很久没有去工作室看望他了。

我真的很幸运，成为吕老师名副其实的学生，他是我的良师益友，更成了我专业道路上的指引者。与吕老师第一次见面记得是在大学三年级的书籍设计课上，吕老师带着我们到香港，和香港理工大学设计学院的老师同学做交流，或聚在教室做有意思的字体重构设计游戏，或跑到室外参观各种设计展。吕老师寓教于乐，令我们这些初涉设计领域的孩子们茅塞顿开。在课堂上，吕老师极力在我们这些年轻人中唤回人之本来具有的与自然亲密接触的创造意识。大学本科期间要准备毕业设计，吕老师带着六个做书籍设计的学生，那时吕老师便发现我的绘画功力还不错，给我找来不少国际一流插画家的作品集让我翻阅学习，在他的鼓励和悉心指点下我完成了人生中的第一本绘本。印刷阶段，吕老师开着当时已不多见的绿色小富康带着我们跑印厂，每个印刷环节都亲自把关，帮助每个人解决遇到的问题。国人普遍的价值认同是"差不离"，但吕老师是分毫不差，他严谨认真的做事态度深深感染了我。

上学期间，我的设计感觉不好，笔头子倒还说得过去，吕老师就一直鼓励我坚持手绘，他拿来各种精美的插图给我学习，让我意识到手绘作品的不可替代的艺术感染力。记得他说过这样一段话："要感谢电子时代，丰富的载体带来百花齐放的可能。正因为数字媒体的发展，作为传达图文主要功能的纸张载体潜力被解放出来。未来图书的个性会日益鲜明，收藏性更强。"这段话同样适用于插图艺术，当今数码插画盛行，手绘纸本插图的独特性则更加凸显了，艺术价值也更高。直到现在我依旧坚持手绘报纸插图和杂志插图，吕老师对书籍设计的这份坚定与专注，一直影响着我，也给了我将手绘插图作为自己毕生职业的那份信心和勇气。

2007年秋，我开始了为期两年半的研究生学业，成了吕老师门下的研究生，他的工作室便成了我的课堂，课程的内容是大量的实践。2008年北京承办夏季奥运会，工作室承担了很多与奥运相关的设计工作。记得2007年冬的一天，吕老师腾出工作室的一间书房，将全开的素描纸贴在白墙上，足足贴了六张，摆上个爬梯，原来他是要交给我一项插图绘制工作，叫我与工作室的伙伴们完成一幅超大的手绘北京旅游地图。想到一下子可以画一面墙，真是把我美得屁颠屁颠的，我们五六个年轻人在一个星期内像打了鸡血似的没日没夜地画，最后把这六张全开的素描画稿从墙上揭下来，组合在一起扫描、上色、印刷，结合信息图标设计成了北京旅游地图。事后想想，正是因为吕老师对我们这些年轻人的这份信任，才有了这次难得的锻炼机会。

之后不久，吕老师拿着一个老历史学家写给孩子们看的历史书手稿给我看，希望我可以把它画成绘本。起初我有点忐忑，怕自己做不好，但最终在吕老师的鼓励和帮助下，我开始埋头于那个别致的小天地——阁楼的斗室里，查找插图资料。我清清楚楚地记得随便从那被书塞得动弹不得的老书架上取下一本书，上面都会贴着已经卷起的黄色便签，上面定是吕老师密密麻麻的读书笔记或者检索记号,他似乎不愿让自己的笔给书留下任何痕迹,小心翼翼地像爱护小孩子一般呵护着它们。每次恩师从德国法兰克福书展回来,都会把第一手的图像资料拷贝给我,将他的见闻感想讲给我听,把他挑选买回的绘本拿给我借阅。吕老师的精一和专注成就了他的高度,他让我觉得人一辈子能认认真真做一件事有多么了不起。2007 年初夏,吕老师还带着我去看望了他的恩师、连环画界的泰斗——贺友直老师。在那间古朴的老房子里,我有幸看到很多珍贵的插图原作,聆听了贺老师关于插图创作的教诲,那时的情景仍历历在目。此后一年半在每个创作环节中,都有吕老师的指引和点拨,这些经历令我无法忘却。

◆ 2008年6月与吕敬人老师看望连环画泰斗、美术教育家贺友直老师

吕老师的观念、心态永远像个年轻人,是一个与新事物新时代同步跟进的人。虽然今年的他已 69 岁高龄,须发皆白,但在我认识的他的同辈中,我不记得有任何一位如他这般,随时敏感、肯定,毫不费力地维持着几十年的好心情、好脾气。在我的记忆中,吕老师绝对是个精力过人的前辈,教书、策展、当评委、弄刊物、著述、学术交流、办班……不知亲手促成多少事情,这样一件接一件地操劳,从不见忙碌相,更未有过半句抱怨。大学期间跟随吕老师出国交流学习,我们这些年轻人都要佩服吕老师的好精力,走起来健步如飞,留下我们一群年轻人一路小跑尾随。

毕业后我去了中国日报社做美编,每次去看望吕老师都是周末,而周末吕老师也是在工作室工作的,因此我们的见面地点自然也就约在工作室。进了门,吕老师准会笑盈盈地迎出来,领我到里屋,边收拾还未来得及合上的书和电脑,边找一块没被书籍占据的空位置招呼我坐下。每次见面除了聊天,我定会带上我近期的插图给他看。吕老师每次都会小心翼翼地接过画,戴上眼镜开始仔细"检阅",这之后定是长时间耐心细致的点评,也是我最期待的环节。吕老师习惯边讲边给我找解决问题的范例,每每这时他总会迅速起身,从摆满书籍的书架上准确定位、取下样书,我便惊讶于他那过人的记忆力。2012 年,吕老师退休了,但他退休后的日程安排得反而比先前更满了,开始担任起《书籍设计》的主编,定期他都会给我快递一期最新的《书籍设计》供我阅读学习,想必老师是希望我一如既往地加强学习,业务上切不可有半点荒废。

我很少给吕老师带礼物,唯有 2012 年吕老师生日时,我亲手给他捏了个不过手指高的小陶塑:一个花白须发的老人坐在那,捧着一本红色的大书,凳子的造型则是"吕"字。当时把这个小得可怜的小礼物有点害臊地递到老师面前时,他很是惊喜,随即一手架起眼镜一手托起小陶塑好一阵端详,看得出吕老师很喜欢。去年,吕老师出了一本书叫《我属猪》,书里收录了学生送的画像或制作的玩偶,每个人都用自己的方式来表达对吕老师的敬爱,没料我送给他的小陶塑也出现在书中,看到我那个小小的陶塑被老师精心收藏着,真的很感动。在吕老师身边学习的这些年,他让我领悟到很多,更教会了我如何做人、做事和对待工作的态度,以及对插图事业的专注与热爱,这是一段铭刻在心的经历,也是一种陶冶与修炼,那的珍贵和幸福,是我一生的福运……

2016 年 3 月写于北京

吕敬人 LU JINGREN

书籍设计师，清华大学美术学院教授，国际平面设计联盟（AGI）成员，中国美术家协会平面设计艺委会副主任，中国出版协会装帧艺术工作委员会副主任，中国艺术研究院设计研究院研究员，敬人设计工作室艺术总监，《书籍设计》丛书主编。曾被评为对中国书籍装帧50年产生影响的十位设计家之一，亚洲十大设计师之一，对新中国书籍60年有杰出贡献的60位编辑之一。获首届华人艺术成就大奖，中国设计事业功勋奖，2016年致造物者·非凡时尚人物奖。

作品曾获国内外多项大奖，其中有德国莱比锡"世界最美的书"奖，全国书籍装帧艺术展览金奖。3次获得中国出版装帧设计政府奖，13次获得"中国最美的书"奖。作品在国内外展出，并被多个国家博物馆、图书馆收藏。2012年在德国奥芬巴赫柯林斯勃博物馆举办"吕敬人书籍设计艺术"个展。2014年担任德国莱比锡"世界最美的书"国际评委。

2016年在韩国坡州举办"法古创新·敬人人敬——传承创造：吕敬人的书籍设计与他的10位弟子展"。2017年在美国旧金山举办"吕敬人的设计"个展。2017年11月在北京今日美术馆举办"书艺问道——吕敬人书籍设计40年展"。

Art Director of Jingren Art Design, Book Designer, Illustrator, Professor of Academy of Arts and Design, Tsinghua University, Member of AGI, Chief editor of Book Design Magazine.
He has been named One of Ten Designers Who Had Significant Influences on Chinese Bookbinding and Design Industry in the Past 50, One of The Top 10 Asian Designer, The Outstanding Contribution to Chinese 60 Years Publishing Enterprise Award, The first Chinese Artistic Achievement Award, Chinese Design Career Award, Outstanding Fashion Personages Award in 2016, etc.
Lu has won several native and international awards. In 2009 and 2012, he won the Appreciate Honnor and the Silver Medal of The Best Designed Book From The World international Competition.
2012, Klingspor-Museum's first Chinese book artist solo exhibition "Delikat: Lu Jingren Book Design" was held in Offenbach, Germany. 2014, he was invited to join the jury panel of "Best Book Design from All Over the World International Competition". The annual book design competition has been held in Leipzig since 1963.
2016, "Imitating and Innovating: Book Design by Lu Jingren and his 10 Proteges" was held in Pajubookcity, Korea.
2017, the book artist solo exhibition "Lu Jingren Book Design" was held in San Francisco, USA.

◆ 2016年10月"法古创新·敬人人敬 —— 传承创造：吕敬人的书籍设计与他的10位弟子展"在韩国坡州书城举行
Oct. 2016, "Imitating and Innovating: Book Design by Lu Jingren and his 10 Proteges" was held in Pajubookcity, Korea

◆ 国际平面设计联盟（AGI）成员参观展览后留念
Alliance Graphique Internationale (AGI) Members visited the exhibition

Respecting Earns Respect

人
敬人
人 敬 人
敬 人 人 敬
敬人人人人敬
人敬人人人人敬人

图书在版编目 (CIP) 数据

法古创新·敬人人敬：吕敬人的书籍设计：汉英对照 / 吕敬人 著 .
—上海：上海人民美术出版社，2018.7
ISBN 978-7-5586-0533-8

Ⅰ . ①法… Ⅱ . ①吕… Ⅲ . ①书籍装帧－设计－汉、英 Ⅳ . ① TS881
中国版本图书馆 CIP 数据核字 (2017) 第 236420 号

法古创新·敬人人敬：
吕敬人的书籍设计

著　　者：	吕敬人
出 品 人：	顾　伟
责任编辑：	张　璎
技术编辑：	程佳华
书籍设计：	typo_d
审　　读：	张柏如
出版发行：	上海人民美术出版社
	（上海长乐路 672 弄 33 号 邮政编码：200040）
印　　刷：	北京富诚彩色印刷有限公司
版　　次：	2018 年 7 月第一版
印　　次：	2018 年 7 月第一次
开　　本：	880×1230mm　1/16
印　　张：	23.25
字　　数：	75 千字
图　　幅：	538
印　　数：	0001－3300 册
书　　号：	ISBN 978-7-5586-0533-8
定　　价：	268.00 元